AF595102

# Optimization of Biodiesel and Biofuel Process

# Optimization of Biodiesel and Biofuel Process

Editors

**Diego Luna**
**Antonio Pineda**
**Rafael Estevez**

MDPI • Basel • Beijing • Wuhan • Barcelona • Belgrade • Manchester • Tokyo • Cluj • Tianjin

*Editors*
Diego Luna
Universidad de Córdoba
Spain

Antonio Pineda
Universidad de Córdoba
Spain

Rafael Estevez
Universidad de Córdoba
Spain

*Editorial Office*
MDPI
St. Alban-Anlage 66
4052 Basel, Switzerland

This is a reprint of articles from the Special Issue published online in the open access journal *Energies* (ISSN 1996-1073) (available at: https://www.mdpi.com/journal/energies/special_issues/Biodiesel_and_Biofuel_Process).

For citation purposes, cite each article independently as indicated on the article page online and as indicated below:

LastName, A.A.; LastName, B.B.; LastName, C.C. Article Title. *Journal Name* **Year**, *Volume Number*, Page Range.

**ISBN 978-3-0365-0278-6 (Hbk)**
**ISBN 978-3-0365-0279-3 (PDF)**

# Contents

# About the Editors

**Diego Luna** is currently a Professor of Organic Chemistry at the University of Cordoba, where he has carried out his research work for forty years in the fields of heterogeneous catalysis, homogeneous-heterogenized catalysis, fine chemistry and biotechnology, dedicating himself during the last decade mainly to the field of biofuels. Throughout these four decades, he has participated in more than 30 research projects, funded by the Spanish Ministry of Research and Science or by the EU. As a result of this research career, he has obtained 10 Invention Patents, and in 2007, he participated in the creation of the technology-based company (spin-off) SENECA GREEN CATALYSTS, SL, which was established to carry out the commercial exploitation of two of these patents, owned by the University of Córdoba and applied to the production of biodiesel. He has participated in several hundred international congresses of his specialty, where he has presented more than 300 communications. He has published 94 book chapters and 192 scientific articles in high-impact international journals. Currently, his research group is involved, among other projects, in "obtaining biofuels, Fine Chemical products and hydrogen through triglyceride biorefinery", for which the concept of biorefinery is applied to the management of oils and fats, preferably reused or inedible, in the paradigm of the so-called circular economy. Currently, he is also an Editorial Board Member of several peer-reviewed open access journals, published monthly online by MDPI, such as *Energies* (https://www.mdpi.com/journal/energies/editors) (ISSN 1996-1073; CODEN: ENERGA), *Catalysts* (ISSN 2073-4344; CODEN: CATACJ) and *Eng* (ISSN x123-x123).

**Antonio Pineda** is a chemist in the Organic Chemistry Department at the University of Cordoba, Spain. He completed his Ph.D. in 2013 at the University of Cordoba, focusing on the study of novel preparation methodologies for heterogeneous catalysts. Currently, he is working on the study of heterogeneous catalysts for biomass valorization processes such as oxidation of lignin derivatives or continuous-flow hydrogenation of platform molecules. He is a member of the Spanish Catalysis Society (SECAT).

**Rafael Estevez** is a postdoctoral researcher in the Organic Department at the University of Cordoba. He completed his Ph.D. in 2017 at the University of Cordoba, focusing on the catalytic valorization of glycerol. He was awarded with the Spanish Catalysis Society (SECAT) Best Doctoral Thesis in his field in 2018. Currently, he is working on heterogeneous catalysis for biomass valorization, focusing on the development of new biofuels from platform molecules such as glycerol and additives for (bio)fuels.

# Preface to "Optimization of Biodiesel and Biofuel Process"

Although the utilization of green hydrogen as an energy vector for the decarbonization of the planet seems to be an irreversible decision, the transition from current energy sources to this new technology requires a period of no less than three or four decades. In addition, the enormous number of transport vehicles currently operating with diesel combustion engines must be considered because they must continue to operate throughout their useful life with diesel fuels or biofuels with similar properties.

To this very high number of vehicles currently in use, those that are currently being built and those with foreseeable construction in the next two or three decades must be added, providing that the deadlines set by the main car, truck, boat, and aircraft builders are met. Therefore, this energy transition needs the contribution of viable technical and economic solutions in order to gradually replace the fossil fuels currently used by several biofuels, both to reduce the levels of $CO_2$ emissions and polluting emissions in cities that seriously affect the health of citizens.

In this Special Issue, a series of papers are collected that try to contribute to this effort, in order to enable the substitution of fossil fuels, in the greatest proportion possible in this transition period, which could be extended even for several decades. Thus, some research works are collected here that provide solutions to make the manufacture of conventional biodiesel more viable or that provide solutions for the reuse of glycerin, which is currently being produced in conventional biodiesel production plants. With glycerin, which is practically a waste, it is proposed to manufacture additives that improve the quality of polluting emissions. Solutions are also provided to try to avoid the production of glycerin as a residue, integrating this molecule directly as a derivative in biofuels. However, the most viable and effective solution at this time would be the direct application of straight vegetable oils in triple mixtures, with fossil diesel and an organic solvent, of a renewable nature such as ethyl ether or ethyl acetate. Even non-renewable gasoline can be used in a triple blend to adapt the viscosity of vegetable oils for use in conventional diesel engines. These triple blends avoid any synthesis process for the production of a biofuel and are capable of making a current diesel engine work correctly, with high levels of substitution of fossil fuels and with a notable reduction in polluting emissions.

**Diego Luna, Antonio Pineda, Rafael Estevez**

*Editors*

*Article*

# *Rhizomucor miehei* Lipase Supported on Inorganic Solids, as Biocatalyst for the Synthesis of Biofuels: Improving the Experimental Conditions by Response Surface Methodology

**Juan Calero [1], Diego Luna [1,*], Carlos Luna [1], Felipa M. Bautista [1], Beatriz Hurtado [1], Antonio A. Romero [1], Alejandro Posadillo [2] and Rafael Estevez [1]**

1 Departamento de Química Orgánica/Universidad de Córdoba, Campus de Rabanales, Ed. Marie Curie, 14014 Córdoba, Spain; p72camaj@gmail.com (J.C.); qo2luduc@uco.es (C.L.); qo1baruf@uco.es (F.M.B.); b.hurtadocebrian@gmail.com (B.H.); qo1rorea@uco.es (A.A.R.); rafa_20_15@hotmail.com (R.E.)

2 Seneca Green Catalyst S.L., Campus de Rabanales, 14014 Córdoba, Spain; seneca@uco.es

* Correspondence: diego.luna@uco.es; Tel.: +34-957212065

Received: 12 February 2019; Accepted: 26 February 2019; Published: 2 March 2019

**Abstract:** Two inorganic solids have been evaluated as supports of Lipozyme RM IM, a *Rhizomucor miehei* lipase immobilized on a macroporous anion exchange resin, in order to improve its application as a biocatalyst in the synthesis of biofuels. The experimental conditions have been optimized to get the selective transesterification of sunflower oil, by using a multi-factorial design based on the response surface methodology (RSM). In this way, the effects of several reaction parameters on the selective ethanolysis of triglycerides to produce Ecodiesel, a biodiesel-like biofuel constitute by one mole of monoglyceride (MG) and two moles of fatty acid ethyl ester (FAEE), have been evaluated. Thus, it was obtained that a 6:1 oil/ethanol molar ratio, 0.215 g of biocatalyst supported in silica-gel (0.015 g Lipase/0.2 g silica-gel), 50 μL of 10 N NaOH, together with previous optimized reaction parameters, 35 °C reaction temperature and 120 min of reaction time, gave the best results (conversions around 70%; selectivity around 65%; kinematic viscosities about 9.3 $mm^2/s$) in the reaction studied. Besides, Lipozyme RM IM, supported on silica-gel, biocatalyst exhibited a very good stability, remaining its activity even after 15 cycles.

**Keywords:** biodiesel; Ecodiesel; selective ethanolysis; sunflower oil; Lipozyme RM IM; *Rhizomucor miehei*; ANOVA method; response surface methodology

---

## 1. Introduction

The transport sector is currently facing an unprecedented challenge. On the one hand, fossil fuels represent more than 95% of the energy employed in this sector, increasing day by day. However, as fossil fuels have a finite nature, they cannot cope with this demand indefinitely. On the other hand, the control of greenhouse gas emissions, including $CO_2$, is mandatory for environmental purposes. Thus, an smooth transition from the current scenario, in which diesel engines work mainly with fossil fuels, to another in which these engines will work mainly with renewable biofuels would allow to respond the increasing demand on fuels, as well as to partially solve the environmental issues [1,2].

In fact, biofuels have been postulated as the best option to replace fossil fuels, since to date, electric motors or vehicles capable of using fuel cells cannot compete yet with explosion or combustion engines, especially in the field of heavy trucks [3], aviation [4,5], or the shipping sector [6]. Therefore, research on renewable fuels capable of replacing fossil fuels and allowing current engines to operate without any modification constitutes a first order priority [7,8]. Among biofuels, biodiesel is considered the best option to replace fossil fuels in compression ignition diesel engines, since no modifications have to be performed [8–11]. Industrial production of biodiesel is currently carried out by homogeneous alkali-catalyzed transesterification of vegetable oils with methanol [12]. After the reaction, biodiesel is repeatedly washed with water to remove glycerol and soaps [13]. In this process, the generation of glycerol as byproduct is a major issue for both, the high amount of glycerol generated (10% by weight of the total biodiesel produced) and the high amount of water employed for removing it [14]. Thus, to solve this drawback, several alternative methods are being investigated.

One option is the production of glycerol derivatives, with rheological properties like methyl esters of fatty acids, during the same transesterification process. This process allows its dissolution in the biofuel and/or in the fossil diesel, increasing the yield of the process and also avoiding the separation of glycerol [15,16]. To do that, methanol is replaced by other donor agents, such as ethyl (or methyl) acetate or dimethyl carbonate. Different catalysts have been investigated, including several lipases [17,18]. As a result of these transesterification reactions, a mixture of three molecules of FAME or FAEE (fatty acid ethyl ester) and one of glycerol carbonate or glycerol triacetate (triacetin) are obtained [19,20].

Other valuable option is the production of a biofuel, with similar physicochemical properties as biodiesel, which integrates glycerol in the form of monoglyceride. In this field, our Research Group has an extensive background, specifically in the production of Ecodiesel, a biofuel obtained by a selective 1,3-regiospecific enzymatic transesterification of the triglycerides (Scheme 1), so that a mixture of two parts of FAEE and one part of MG is obtained [21]. The experimental conditions of the enzymatic process to produce Ecodiesel are much milder than those required for a conventional homogenous process. In addition, the atomic efficiency of the process is very high, since the yield of the process is practically 100%, avoiding, then, any process to eliminate impurities from the biofuel obtained. Last but not least, as a reagent, ethanol is used, a cheaper compound than dimethyl carbonate or ethyl acetate, both described as alternatives for obtaining biofuels that integrate glycerin as soluble compounds in the biofuel.

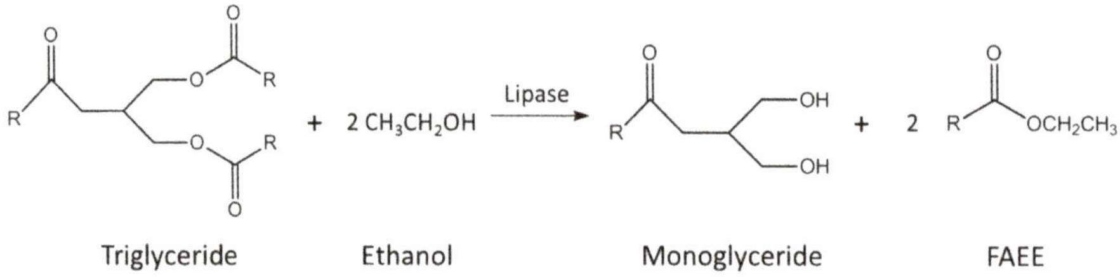

**Scheme 1.** Ecodiesel synthesis by 1,3-selective ethanolysis of triglycerides, which integrates the glycerol as monoglyceride.

The Ecodiesel synthesis was patented using a pig pancreatic lipase PPL [22], although remarkable results have been also obtained over different microbial lipases [23,24]. To optimize the different reaction conditions, such as the effect of temperature, the pH, the oil/alcohol molar ratio, etc. statistical analysis of variance (ANOVA) and response surface methodology (RSM) have been previously employed in Ecodiesel production [25].

Anyhow, despite lipases can be employed as catalyst to perform the Ecodiesel synthesis, their high production cost is still a huge limitation for their use at industrial scale. To overcome this, lipases must be able to be reused in subsequent reactions, being a viable option its immobilization on a support. In this sense, we have recently reported the immobilization of Lipozyme RM IM, a *Rhizomucor miehei* lipase, on a macroporous anion exchange resin and its use as heterogeneous biocatalyst in the selective transesterification of sunflower oil with ethanol [26]. Despite the good results obtained, the low density of the polymer resin lipase makes its recovery very difficult, needing a centrifugation operation at 3500 rpm for 5 minutes. Furthermore, a maximum of six reuses can be attained before the loss of the activity. From an economical point of view, six reuses are not enough if we take under consideration the high price of the lipase.

Hence, in this research, two different aspects have been addressed. On one hand, to avoid the centrifugation process for recovering the biocatalyst, several immobilization methods on inorganic supports have been evaluated. For that purpose, two different cheap inorganic materials, sepiolite and silica, have been evaluated as supports using covalent immobilization and/or physical adsorption.

On the other hand, several reaction parameters have been studied by ANOVA and RSM, to have better insights into the production of Ecodiesel over the *Rhizomucor miehei* lipase. We consider this statistical method very helpful because, as we previously observed employing other enzymes, during the production of Ecodiesel, the properties of the enzymes change depending on these reaction parameters, influencing the measured reaction rates [21,25,27]. In this study, we have evaluated the influence of the amount of lipase, the amount of 10 N NaOH and the oil/ethanol molar ratio in the production of Ecodiesel.

## 2. Materials and Methods

### 2.1. Materials

Sunflower oil for food use was obtained from a local market. Its standard fatty acids profile is: 63.5% linoleic acid, 24% oleic acid, 6.5% palmitic acid, 5% stearic acid and 2% of palmitoleic, with minor amounts of linolenic, behenic, and cetoleic acids. It exhibits a kinematic viscosity value of 32 $mm^2/s$. The water content as determined by the Karl Fisher method was <0.08% and acidity degree 0.2%, expressed as oleic acid content. The palmitic acid, stearic acid, oleic acid, linoleic acid and linolenic acid methyl esters used as standards were obtained from AccuStandard, Inc. 125 Market Street, New Haven, CT 06513, USA and methyl heptadecanoate was purchase from Sigma–Aldrich, San Luis, Misuri, Estados Unidos. All of them were chromatographically pure. Absolute ethanol and sodium hydroxide pure analytical compounds (99%) used were purchased from Panreac, Carrer del Garraf, 2, 08211 Castellar del Vallès, Barcelona, Spain. The Lipozyme RM IM, a *Rhizomucor miehei* lipase immobilized in beads from macroporous anion exchange resins was kindly provided by Novozymes A/S, Krogshøjvej 36, 2880 Bagsværd, Denmark.

### 2.2. Immobilization of Rhizomucor Miehei Lipase on Inorganic Supports

#### 2.2.1. Immobilization of Lipozyme RM IM by Physical Adsorption

The physical adsorption of Lipozyme RM IM has been studied on two different inorganic supports, a natural sepiolite (Tolsa S.A., Zaragoza, Spain) and a commercial silica gel. The sepiolite is a cheap natural silicate with high surface area (226 $m^2/g$) and a fibrous structure. The theoretical formula of the unit cell is $Si_{12}O_{30}Mg_8(OH)_6(H_2O)_4{\cdot}8H_2O$, where the $Si^{4+}$ and the $Mg^{2+}$ can be partially substituted by $Al^{3+}$, $Fe^{2+}$ and alkaline ions. Each Mg atom completes its coordination with two molecules of water (Figure 1). The physical adsorption of lipases on sepiolite requires a previous acid demineralization step, in which the different metal hydroxides, i.e., Al, Fe, alkaline ions and mostly Mg, are extracted [28]. In this case, 40 g of the sepiolite was stirred at RT with a 1M solution of hydrochloric acid (HCl), until no presence of magnesium is detected (24 h).

Afterwards, the channels of the sepiolite can be filled with the lipase, producing its immobilization by physical adsorption. For its part, the immobilization on a silica gel does not require any activation treatment of the silica. Thus, the physical immobilization was carried out according to the following procedure. In a 25 mL round bottom flask, 0.2 g support, 0.01 g of Lipozyme RM IM lipase and 3.5 mL of absolute ethanol are mixed and stirred for 30 min at 700 rpm and 35 °C. As a matter of density of the solids, the corresponding amount of demineralized sepiolite employed as inorganic support was 1.0 g. The biocatalysts obtained by immobilization will be denoted as Lipo-Sep and Lipo-silica, either if Lipozyme was adsorbed on demineralized sepiolite or on silica-gel.

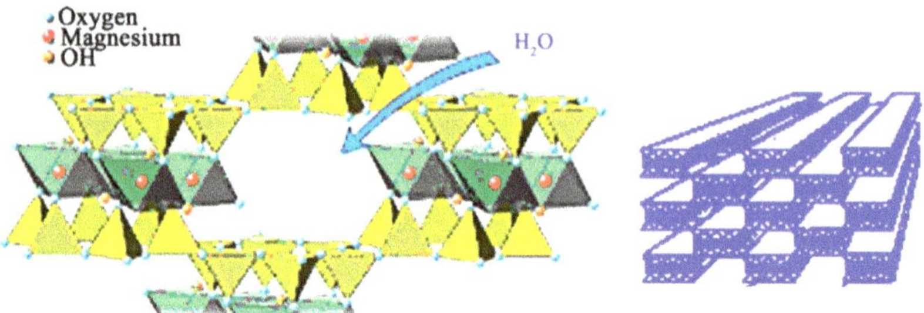


**Figure 1.** Structure of hydrated natural sepiolite.

### 2.2.2. Covalent Immobilization of Lipozyme RM IM on Sepiolite

Analogously to the physical adsorption, sepiolite cannot be directly employed as support for covalent immobilization of lipases. First of all, it has to be subjected to a surface activation process by sol-gel precipitation of $AlPO_4$ on powdered solid sepiolite, in a proportion Sepiolite/$AlPO_4$ 80/20 [24]. Then, two different linkers, *p*-hydroxybenzaldehyde or 4-aminobenzylamine, were employed to interact with the Bronsted acid sites of the support, following a reported procedure [28], as it is shown in Schemes 2a and 3a,b. If 4-aminobenzylamine is employed to modify the functional groups in the surface, tereftaldicarboxaldehyde is also added to react with 4-aminobenzylamine and form the imines bond. Briefly, the immobilization of Lipozyme was carried out at room temperature, by introducing the functionalized inorganic solid (0.2 g) with the Lipozyme RM IM (0.01 g) in a reaction flask with 6 mL of ethanol, stirring in a refrigerator for 24 h. Finally, prior to its use, ethanol (6 mL) was added to the mixture and the solid, with the immobilized Lipozyme was separated by centrifugation. These supports will be denoted as 1-Sep/$AlPO_4$ for the support modified with *p*-hydroxybenzaldehyde and 2-Sep/$AlPO_4$ for the support modified with 4-aminobenzylamine and tereftaldicarboxaldehyde. Finally, the covalent immobilization of the lipase on the modified amorphous Sepiolite/$AlPO_4$ can be achieved by chemical interaction of the organic groups available in the lipase either with the amino group or the aldehyde group, Schemes 2b and 3c. The final biocatalysts will be denoted as Lipo-1Sep/$AlPO_4$ and Lipo-2Sep/$AlPO_4$.

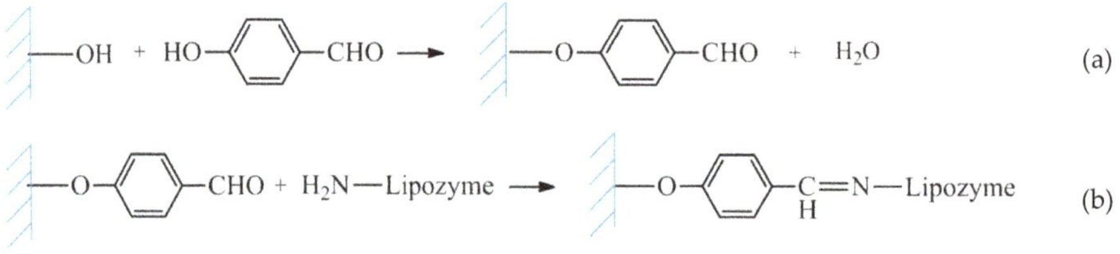

—OH + HO—C₆H₄—CHO → —O—C₆H₄—CHO + $H_2O$ (a)

—O—C₆H₄—CHO + $H_2N$—Lipozyme → —O—C₆H₄—C(H)=N—Lipozyme (b)

**Scheme 2.** Solid phase synthesis, step by step, for the immobilization of the enzyme through the ε-amino groups of the lysine residues. In Step 1, surface OH groups of the supported Sepiolite/$AlPO_4$ are activated by microwave heating with p-hydroxybenzaldehyde; before covalent immobilization of the enzyme through the ε-amino group of lysine residues (step 2).

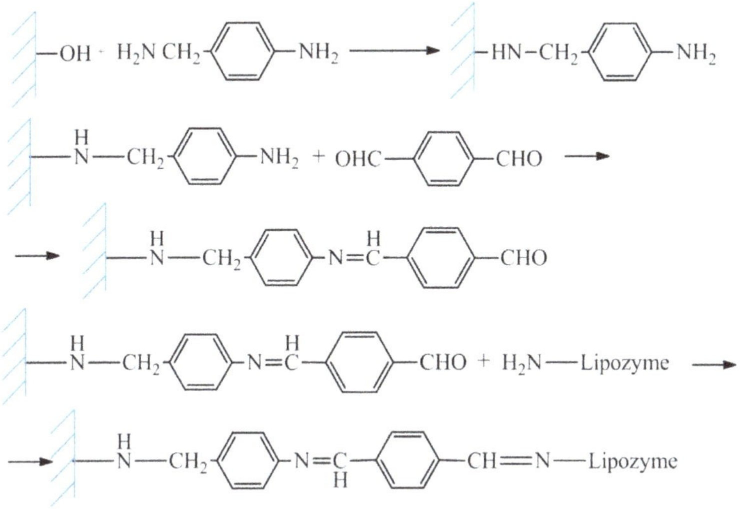

—OH + $H_2N\,CH_2$–C₆H₄–$NH_2$ ⟶ —HN–$CH_2$–C₆H₄–$NH_2$ a)

—NH–$CH_2$–C₆H₄–$NH_2$ + OHC–C₆H₄–CHO ⟶

⟶ —NH–$CH_2$–C₆H₄–N=CH–C₆H₄–CHO b)

—NH–$CH_2$–C₆H₄–N=CH–C₆H₄–CHO + $H_2N$—Lipozyme ⟶

⟶ —NH–$CH_2$–C₆H₄–N=CH–C₆H₄–CH=N—Lipozyme c)

**Scheme 3.** Solid phase synthesis, step by step, for the immobilization of the enzyme through the ε-amino groups of the lysine residues. In Step 1, surface OH groups of the supported $AlPO_4$/sepiolite are activated by microwave heating with 4-aminobenzylamine; In Step 2, tereftaldicarboxaldehyde is reacted through imines bonds also obtained by microwave heating; Finally, in Step 3 covalent immobilization of the enzyme is obtained through the ε-amino group of lysine residues. It must be clarified that water is formed in steps (**a**)–(**c**).

Figure 2 shows the inorganic supports employed to immobilize the Lipozyme RM IM. As can be seen, despite the fact 1 g of sepiolite was employed instead of 0.2 g of the rest of the solids, the final volume was similar for all the supports.

**Figure 2.** Inorganic supports used to immobilize Lipozyme RM IM: (**a**) demineralized sepiolite, (**b**) silica gel, (**c**) $AlPO_4$/sepiolite activated with 4-aminobenzylamine and terephthalato dicarboxaldehyde (**d**) $AlPO_4$/sepiolite activated with *p*-hydroxybenzaldehyde.

### 2.3. Ethanolisis Reactions

The transesterification reactions of sunflower oil were carried out according to the experimental procedure previously described [26]. Briefly, a 25 mL volume round bottom flask was immersed in a thermostatic bath in which temperature and stirring speed were controlled. The temperature was set to 35 °C whereas the stirring speed was set at 700 rpm to avoid mass transfer limitations. In a typical run, 9.4 g (12 mL, 0.01 mol) of sunflower oil was introduce in the batch reactor together with variable oil/ethanol volume ratio, corresponding to a molar ratio between 1/4 and 1/6, different amounts of supported Lipozyme RM and also different amounts of NaOH 10N (0–75 µL). Blank experiments in presence of the highest quantity of NaOH 10N solution were performed to rule out a potential contribution from the homogeneous NaOH catalyzed reaction. In these blank experiments, less than 10% of conversion was obtained, indicating that the homogenous catalysis contribution is negligible under the investigated conditions. At the end of the reaction, the biocatalyst was employed without any treatment, in order to simulate the operational conditions employed in the industry. Thus, by simple decanting for half an hour, the biocatalyst is maintained at the bottom of the flask and the product is retired. Then, a new charge of reactants is added and the reaction is launched again.

### 2.4. *Analytical Method*

Reaction products were monitored by capillary column gas chromatography, using a 5890 series II instrument (Hewlett-Packard, Palo Alto, California, USA) equipped with a flame ionization detector (FID) and an HT5 capillary column (25 m × 0.32 mm × 0.10 µm), using *n*-hexadecane (cetane) as internal standard. The heating ramp was: 50 to 200 °C at a rate of 7 °C/min, followed by another ramp from 200 to 360 °C at a rate of 15 °C/min, maintaining the oven temperature at 360 °C for 10 min [29]. The quantification of the ethyl esters (FAEE), monoglycerides (MG), diglycerides (DG) and triglycerides (TG) allow to determine the Conversion, i.e., Conv = FAEE + MG + DG, whereas the Selectivity is calculated as the sum of FAEE + MG. The differentiation between both parameters is determined by the similarity of the FAEE and MG with the standard n-hexadecane. Therefore, it represents the proportion of biofuel directly comparable with fossil fuels.

### 2.5. *Determination of Kinematic Viscosity*

The kinematic viscosity has been measured in an Ostwald-Cannon-Fenske capillary viscometer (Proton Routine Viscometer 33200, size 150, Proton Technology AB, Sjöåkravägen 28, SE-564 31 Bankeryd, Sweden), determining the time required for a certain volume of liquid to pass between two marked points on the instrument, placed in an upright position. From the flow time (t), expressed in seconds, we obtain the kinematic viscosity expressed in centistokes, $\upsilon = C \cdot t$, where C is the calibration constant of the measurement system in $mm^2/s^2$, which is specified by the manufacturer (0.040350 $mm^2/s^2$ at 40 °C, in this case). All measures have been carried out in duplicate and are presented as the average of both, proving that the variation is below 0.35% between measures, as required by the standard American Society for Testing and Materials ASTM D2270-79 method for the calculating viscosity index from kinematic viscosity at 40 and 100 °C.

### 2.6. *Experimental Design*

The effect of process parameters in the enzymatic transesterification reaction and the optimum conditions for the selectivity and viscosity were studied using a multifactorial design of experiments with three factors run by the software StatGraphics®version XVI centurion; Statgraphics.Net, C/Bravo Murillo 350, 1°, 28020 Madrid, Spain. Two of the variables were studied at three levels and the other one at two levels (Table 1), giving us 36 runs. The experiments were performed in random order. The experimental parameters selected for this study were immobilized lipase amount, oil/ethanol molar ratio and NaOH 10N amount. Table 1 shows the coded and actual values of the process parameters used in the design matrix.

**Table 1.** Process parameters in factorial design: coded and actual values.

| Variables | Unit | Levels | | |
|---|---|---|---|---|
| | | −1 | 0 | 1 |
| Immobilized Lipozyme RM IM | g | 0.01 | 0.02 | 0.03 |
| Oil/ethanol ratio (*v*/*v*) | mL/mL | 12/2.9 | - | 12/3.5 |
| NaOH 10N amount | μL | 25 | 37.5 | 50 |

As can be seen in Table 1, the ranges studied are the following: weight of Lipozyme RM IM, from 0.01 to 0.03 g, oil/ethanol volume ratio between 12/2.9 to 12/3.5 mL/mL (equivalent to molar ratios from 1/4 to 1/6, approximately) and the amount of 10N NaOH between 25 and 50 μL. Each experiment is done in triplicate to ensure about the reproducibility of each reaction and improve the model.

### 2.7. Statistical Analysis

The experimental data obtained from experimental design were analyzed by RSM [29,30]. A mathematical model, following a second-order polynomial equation, was developed to describe the relationships between the predicted response variable (selectivity or viscosity) and the independent variables of reaction conditions, as shown in Equation (1), where y is the predicted response variable; $\beta_0$, $\beta i$, $\beta ii$, $\beta ij$ are the intercept, linear, quadratic and interaction constant coefficients of the model, respectively; $Xi$, $Xj$ ($i$ = 1, 3; $j$ = 1, 3; $i \neq j$) represent the coded independent variables (reaction conditions):

$$y = \beta_0 + \sum_{i=1}^{3} \beta_i X_i + \sum_{i=1}^{3} \beta_{ii} {X_i}^2 + \sum \sum_{i<j=1}^{3} \beta_{ij} X_i X_j \quad (1)$$

Response surface plots were developed using the fitted quadratic polynomial equation obtained from regression analysis, holding one of the independent variables at constant values corresponding to the stationary point and changing the order two variables. The quality of the fit of the polynomial model equation was evaluated by the coefficient of determination $R^2$. Likewise, its regression coefficient significance was checked with F-test. Confirmatory experiments were carried out to validate the model, using combinations of independent variables which were not part of the original experimental design, but included in the experimental region.

## 3. Results and Discussions

### 3.1. Immobilization of the Commercial Biocatalyst Lipozyme RM IM on an Inorganic Support

Table 2 shows the conversion and selectivity values obtained in the ethanolysis reaction with the Lipozyme RM IM over all the different supports here studied, as well as the viscosity of the biofuel synthesized.

**Table 2.** Ethanolysis reactions with Lipozyme RM IM immobilized on different inorganic supports, under the standard experimental reaction conditions: 12 mL of sunflower oil, 3.5 mL of absolute ethanol, at a temperature of 35 °C, 25 μL of 10 N aqueous solution of NaOH, 0.01 g of Lipozyme RM IM, 2 h reaction times and an agitation of 700 rpm, and the indicated amounts of support.

| Support | Viscosity (cSt) | Conversion (%) | Selectivity (%) |
|---|---|---|---|
| Unsupported Lipozyme | 7.9 | 100 | 62.4 |
| Lipo-Sep | 18.8 | - | - |
| Lipo-Silica | 11.6 | 83.0 | 31.4 |
| Lipo-1Sep/$AlPO_4$ | 23.7 | [1] - | [1] - |
| Lipo-2Sep/$AlPO_4$ | 16.2 | [1] - | [1] - |

[1] The two initial phases of the reaction mixture (oil and ethanol) are kept separate, an unambiguous sign of inactivity in the reaction.

As can be seen, the covalent immobilization of the Lipozyme RM IM on Sepiolite support, independently on the procedure employed, produce its complete deactivation. It seems that the presence of functional groups that make possible the covalent immobilization of the lipase, cause the deactivation of its active site somehow. Likewise, the adsorption of the Lipozyme on the demineralized sepiolite also caused the deactivation of the lipase. However, if the Lipozyme is physically adsorbed on silica-gel, reasonably good results are obtained. This immobilization would be well-founded by a physical interaction between the microporous anion exchange resins, in which the lipases is already immobilized, and the silica support. Thus, the interaction allowed increase the low density of the organic polymer without affect the actives sites of the lipase. In order to corroborate the economic feasibility of the lipase immobilization procedure, the performance of the lipase after different reuses must be evaluated. Figure 3 shows the conversion, selectivity and the viscosity of the biofuel obtained, after 15 successive reactions, over the Lipo-silica biocatalyst. As can be seen, the stability of the Lipo-silica is outstanding. In fact, even after the 15th reuse, the conversion and the selectivity are pretty similar to that obtained after the first use. Furthermore, the viscosity of the biofuel obtained is the same for every reaction (10.5 ± 0.5 cSt). These results are very encouraging, above all if we compare it with those obtained with the Lipozyme without immobilize, with which only six reuses before a total loss of the activity is observed [26]. Furthermore, considering the operative aspects at plant scale, the Lipo-silica catalyst only requires that the product of the reaction be extracted from the reactor and then, a new charge of reactants can be added for starting the subsequent reaction. For its part, the Lipozyme without immobilization requires a centrifugation process, much more difficult from an operational point of view.

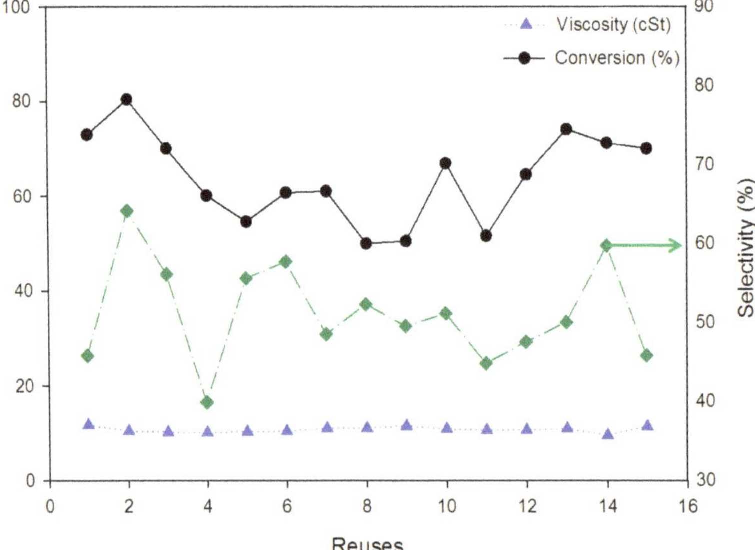


**Figure 3.** Reuse of the *Rhizomucor miehei* system of Novozyme (Lipozyme RM IM) immobilized with silica gel (Lipo-silica), by simple physical attraction, operating under the standard conditions indicated in Table 2.

### *3.2. Analysis of Variance (ANOVA) and Optimization of the Reaction Parameters by RSM*

Given the good behavior of the Lipo-silica biocatalytic system in the Ecodiesel production, a multilevel factorial experimental design has been carried out in order to analyze the effects of experimental parameters in the enzymatic selective ethanolysis reaction of sunflower oil. In this sense, on the basis of previous researches [27], the weight of Lipo-silica, the oil/ethanol volume ratio and the amount of 10 N NaOH have been selected as the most relevant parameters. To evaluate the magnitude of these parameters on the selectivity of the partial ethanolysis reaction of sunflower oil, some limits have been set for each variable, as indicated in Table 1. It is also intended to determine the influence of these variables on the kinematic viscosity of the final mixture, given that the viscosity is the parameter which determines whether the biofuel obtained can be employed in current engines. Therefore, the experimental design consists of 18 experiments (runs) carried out in duplicate and in a random way to minimize possible errors. The sequence of the performed experiments is shown in Table 3.

**Table 3.** Experiments matrix of factorial design and the response obtained for viscosity and selectivity.

| Run | Studied Variables | | | Output Variables | |
|---|---|---|---|---|---|
| | Amount Lipozyme RM IM (g) | Oíl/ethanol ratio (mol/mol) | Amount of 10 N NaOH (μL) | Viscosity (cSt) | Selectivity (%) |
| 1 | 1 | −1 | 0 | 11.36 | 39.7 |
| 2 | 1 | 1 | −1 | 10.16 | 50.1 |
| 3 | −1 | 1 | 1 | 9.98 | 67.9 |
| 4 | 0 | 1 | −1 | 9.88 | 49.8 |
| 5 | −1 | −1 | 0 | 11.48 | 45.3 |
| 6 | 1 | −1 | 1 | 11.56 | 47.6 |
| 7 | −1 | 1 | −1 | 10.56 | 51.4 |
| 8 | −1 | 1 | 0 | 9.09 | 55.2 |
| 9 | 0 | −1 | −1 | 11.84 | 40.7 |
| 10 | 0 | −1 | 0 | 11.52 | 41.1 |
| 11 | 0 | 1 | 1 | 8.92 | 69.8 |
| 12 | 1 | 1 | 0 | 10.14 | 51.6 |
| 13 | −1 | −1 | 1 | 10.82 | 52.7 |
| 14 | 0 | 1 | 0 | 9.31 | 56.2 |
| 15 | 0 | −1 | 1 | 9.98 | 48.7 |
| 16 | 1 | 1 | 1 | 9.86 | 67.1 |
| 17 | 1 | −1 | −1 | 11.64 | 35.1 |
| 18 | −1 | −1 | −1 | 12.29 | 43.2 |
| | | Repeated experiments | | | |
| 19 | 1 | 1 | 1 | 9.94 | 58.7 |
| 20 | 0 | −1 | 1 | 9.94 | 53.1 |
| 21 | 1 | −1 | 1 | 11.09 | 44.4 |
| 22 | −1 | 1 | 1 | 10.18 | 64.5 |
| 23 | 1 | −1 | −1 | 12.21 | 34.5 |
| 24 | −1 | 1 | 0 | 10.01 | 60.5 |
| 25 | −1 | −1 | 0 | 11.52 | 45.7 |
| 26 | −1 | 1 | −1 | 10.01 | 54.7 |
| 27 | 0 | 1 | −1 | 8.89 | 56.6 |
| 28 | −1 | −1 | −1 | 11.56 | 36.8 |
| 29 | 0 | 1 | 1 | 9.53 | 61.8 |
| 30 | 0 | 1 | 0 | 9.73 | 59.7 |
| 31 | 0 | −1 | 0 | 10.44 | 44.3 |
| 32 | 1 | 1 | 0 | 10.88 | 54.3 |
| 33 | 0 | −1 | −1 | 11.26 | 33.2 |
| 34 | 1 | −1 | 0 | 11.48 | 37.4 |
| 35 | −1 | −1 | 1 | 10.26 | 50.2 |
| 36 | 1 | 1 | −1 | 10.03 | 50.1 |

From these data, we are able to determine the correlation between the experimental variables here studied with the output variables (selectivity and kinematic viscosity), by using the Statgraphics software and a multivariate statistical analysis (ANOVA). As can be seen in Tables 4 and 5, the quadratic polynomial model was highly significant, allowing us to understand which variables have a greater impact on the selectivity and kinematic viscosity.

**Table 4.** Analysis of variance (ANOVA) for selectivity.

| Source | Sum of Squares | Freedom Degree | Mean Square | F-Ratio | p-Value |
|---|---|---|---|---|---|
| A: Amount Lipozyme RM | 137.76 | 1 | 137.76 | 17.05 | 0.0003 |
| B: Ratio oil/Ethanol | 1969.88 | 1 | 1969.88 | 243.81 | 0.0000 |
| C: Amount NaOH 10N | 941.254 | 1 | 941.254 | 116.50 | 0.0000 |
| AA | 13.6068 | 1 | 13.6068 | 1.68 | 0.2058 |
| AB | 6.93375 | 1 | 6.93375 | 0.86 | 0.3628 |
| AC | 0.09 | 1 | 0.09 | 0.01 | 0.9168 |
| BC | 0.63375 | 1 | 0.63375 | 0.08 | 0.7816 |
| CC | 23.0068 | 1 | 23.0068 | 2.85 | 0.1035 |
| blocks | 4.48028 | 1 | 4.48028 | 0.55 | 0.4632 |
| Total error | 210.071 | 26 | 8.07964 | | |
| Total (corr.) | 3307.72 | 35 | | | |

$R^2 = 0.849$; $R^2(\text{Adj.}) = 0.805$.

**Table 5.** Analysis of variance (ANOVA) for viscosity.

| Source | Sum of Squares | Freedom Degree | Mean Square | F-Ratio | p-Value |
|---|---|---|---|---|---|
| A: Amount Lipozyme RM | 0.279504 | 1 | 0.279504 | 1.59 | 0.2189 |
| B: Ratio oil/Ethanol | 17.5701 | 1 | 17.5701 | 99.77 | 0.0000 |
| C: Amount NaOH 10N | 2.8497 | 1 | 2.8497 | 16.18 | 0.0004 |
| AA | 3.39301 | 1 | 3.39301 | 19.27 | 0.0002 |
| AB | 0.00220417 | 1 | 0.00220417 | 0.01 | 0.9118 |
| AC | 0.158006 | 1 | 0.158006 | 0.90 | 0.3523 |
| BC | 1.51504 | 1 | 1.51504 | 8.60 | 0.0069 |
| CC | 0.0325125 | 1 | 0.0325125 | 0.18 | 0.6710 |
| Blocks | 0.0568028 | 1 | 0.0568028 | 0.32 | 0.5750 |
| Total error | 4.57882 | 26 | 0.176109 | | |
| Total (corr.) | 30.4357 | 35 | | | |

$R^2 = 0.936$; $R^2(\text{Adj.}) = 0.917$.

The correlation coefficients values, $R^2$, were 0.849 and 0.936 for selectivity and kinematic viscosity, respectively. These results imply a good fit between models and experimental data in Pareto graphics (Figure 4). Furthermore, the adjusted correlation coefficients $R^2$(Adj) were 0.805 for selectivity and 0.917 for selectivity and kinematic viscosity (Tables 4 and 5). The results here collected pointed out that all the parameters studied has an important influence on the selectivity of the process, specially the oil/ethanol ratio ($v/v$) and. However, according to the kinematic viscosity, the most relevant parameters are the oil/ethanol ratio ($v/v$) and the amount of 10 N NaOH employed ($p < 0.05$).

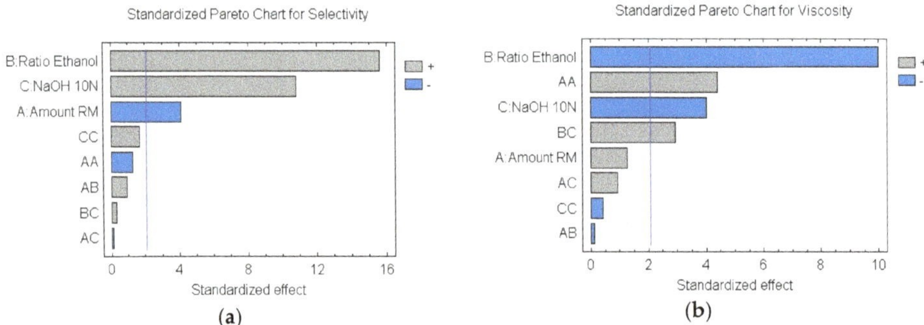


**Figure 4.** Pareto graphics: (**a**) for selectivity and (**b**) for viscosity.

Likewise, if we represent the influence of the different factors on selectivity and kinematic viscosity (Figure 5), it is easy to deduce that the higher the positive slope, the higher the effect of the factor. Thus, it can be seen in Figure 5a,b, how the oil/ethanol ratio is the parameter with the highest statistical significance not only on the selectivity but also on the kinematic viscosity.

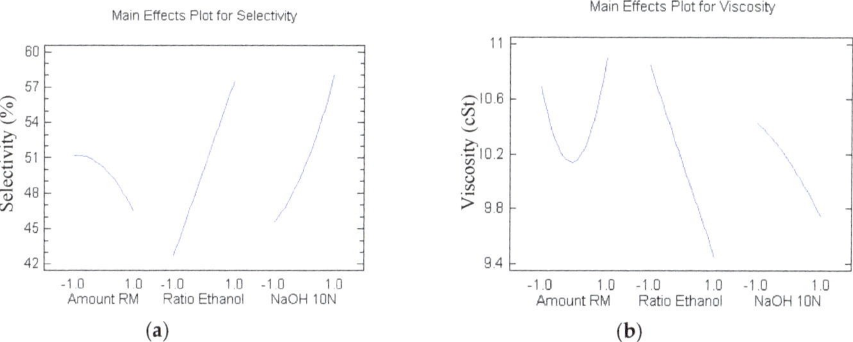


**Figure 5.** Graphical representation of the different factors on: (**a**) selectivity; (**b**) on the viscosity.

Furthermore, the software also allows obtaining equations, after the elimination of non-influent parameters in the model, for selectivity and kinematic viscosity, whose $R^2$ values were 0,741 and 0.873, respectively. The equations obtained were quite simpler as compared to initial ones. These equations describe the created model and gives solutions for the dependent variable based on the independent variable combinations, whether they are or not significant in the response. Thus, considering that R is the Oil/Ethanol ratio ($v/v$); N is the amount of 10 N NaOH in μL and A, amount of Lipozyme RM, we have the following equations:

$$\text{Selectivity} = 50.1194 + 7.39722\text{R} + 0.1625\text{RN} - 1.30417\text{A}^2 \quad (2)$$

$$\text{Viscosity} = 10.1458 + 0.107917\text{A} - 0.698611\text{R} + 0.65125\text{A}^2 \quad (3)$$

The surface plots described by the regression model were drawn to display the effects of the independent variables on selectivity and kinematic viscosity, Figure 6. This model shows that the optimum values for the parameters to maximize selectivity were: low lipase amount (0.012 g), high oil/ethanol molar ratio (1/6) and high 10 N NaOH amount (50 μL). Regarding the optimum values to maximize the viscosity: lipase amount (0.017 g), which is a bit higher than that for selectivity, oil/ethanol molar ratio (1/6) and amount of 10 N NaOH (50 μL). Thus, a selectivity value around 66% and viscosity around 9.5 cSt can be achieved under these conditions. The biofuel obtained at these conditions is feasible to be employed in a mix with diesel fuel in blends of 50%, or even more, falling within the acceptance limits of the EN 14214.

To validate the proposed models, a serie of experiments, whose conditions are among the range of the studied variables, has been carried out. Thus, the experimental values of selectivity and kinematic viscosity have been collected in Table 6. Furthermore, a comparison between the experimental values and the theoretical ones has been also compiled. As can be seen in Figure 7, a good adjustment of the model, with values of $R^2$ of 84.955%, for Selectivity and 93.649% for Viscosity, has been obtained, corroborating the good correlation between the experimental values and those predicted by the model.

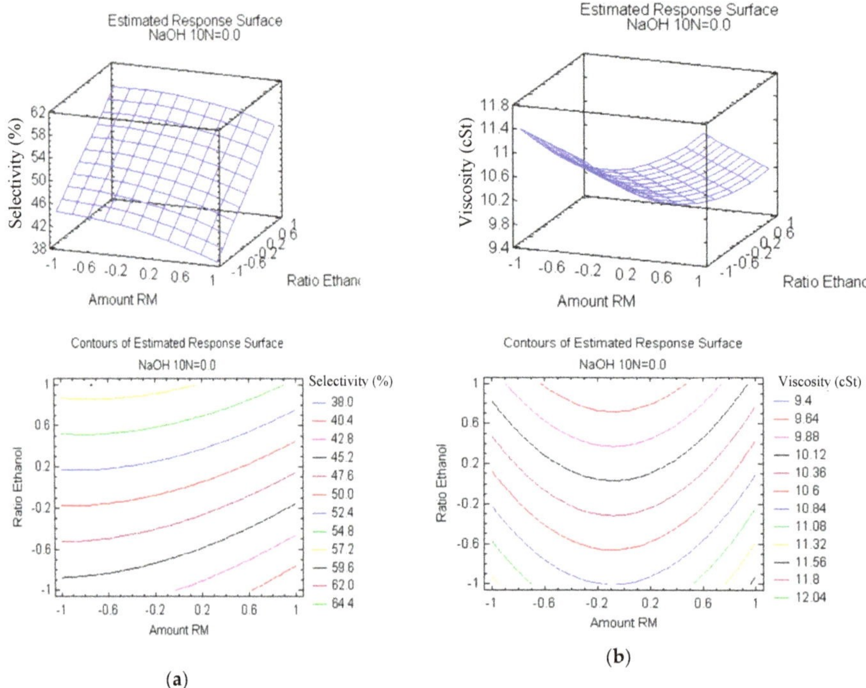


(a) (b)

**Figure 6.** Graphic representation of the different factors on: (**a**) selectivity; (**b**) viscosity.

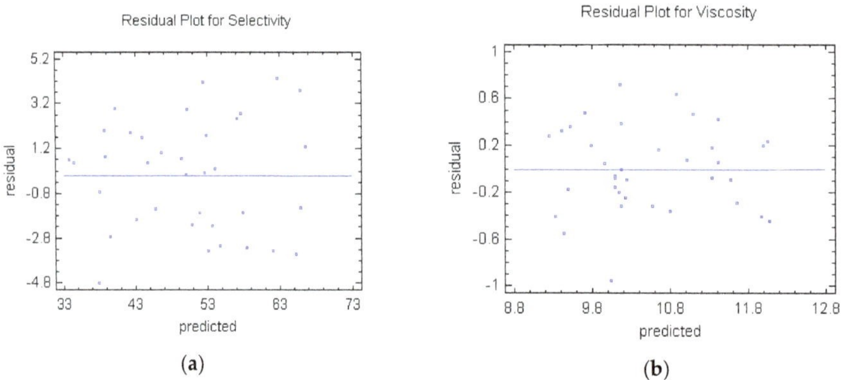


(**a**) (**b**)

**Figure 7.** Probability graphs of (**a**) Selectivity and (**b**) Viscosity.

Therefore, it can be concluded that the model created to correlate the reaction parameters (amount of lipase, ratio oil/ethanol and amount of NaOH 10N aqueous solution) with the selectivity and the kinematic viscosity is acceptable, since it is able to explain the variability produced in these experimental parameters.

**Table 6.** Experiments matrix of factorial design and the response obtained for selectivity and viscosity. Experimental values obtained for each reaction of the experimental design for the Selectivity and Kinematic Viscosity at 40 °C, collected in Table 3, compared to the values obtained when applying the theoretical model proposed.

| Run | Studied Variables | | | Selectivity (%) | | Viscosity (cSt) | |
|---|---|---|---|---|---|---|---|
| | Amount Lipozyme RM IM (g) | Oíl/ethanol Ratio | Amount of 10 N NaOH | Exp. Value | Predicted | Exp. Value | Predicted |
| 1 | 1 | −1 | 0 | 39.7 | 38.8375 | 11.36 | 11.6529 |
| 2 | 1 | 1 | −1 | 50.1 | 50.0528 | 10.16 | 10.1667 |
| 3 | −1 | 1 | 1 | 67.9 | 66.6194 | 9.98 | 9.7834 |
| 4 | 0 | 1 | −1 | 49.8 | 53.1403 | 9.88 | 9.51653 |
| 5 | −1 | −1 | 0 | 45.3 | 44.7042 | 11.48 | 11.4179 |
| 6 | 1 | −1 | 1 | 47.6 | 46.5583 | 11.56 | 11.0927 |
| 7 | −1 | 1 | −1 | 51.4 | 53.6194 | 10.56 | 10.1688 |
| 8 | −1 | 1 | 0 | 55.2 | 58.4236 | 9.09 | 10.0399 |
| 9 | 0 | −1 | −1 | 40.7 | 38.6708 | 11.84 | 11.4162 |
| 10 | 0 | −1 | 0 | 41.1 | 43.075 | 11.52 | 10.8842 |
| 11 | 0 | 1 | 1 | 69.8 | 65.9903 | 8.92 | 9.32986 |
| 12 | 1 | 1 | 0 | 51.6 | 54.7069 | 10.14 | 10.2365 |
| 13 | −1 | −1 | 1 | 52.7 | 52.575 | 10.82 | 10.659 |
| 14 | 0 | 1 | 0 | 56.2 | 57.8694 | 9.31 | 9.48694 |
| 15 | 0 | −1 | 1 | 48.7 | 50.8708 | 9.98 | 10.2246 |
| 16 | 1 | 1 | 1 | 67.1 | 62.7528 | 9.86 | 10.1788 |
| 17 | 1 | −1 | −1 | 35.1 | 34.5083 | 11.64 | 12.0856 |
| 18 | −1 | −1 | −1 | 43.2 | 40.225 | 12.29 | 12.0494 |
| | Repeated experiments | | | | | | |
| 19 | 1 | 1 | 1 | 58.7 | 62.0472 | 9.94 | 10.0994 |
| 20 | 0 | −1 | 1 | 53.1 | 50.1653 | 9.94 | 10.1451 |
| 21 | 1 | −1 | 1 | 44.4 | 45.8528 | 11.09 | 11.0133 |
| 22 | −1 | 1 | 1 | 64.5 | 65.9139 | 10.18 | 9.70396 |
| 23 | 1 | −1 | −1 | 34.5 | 33.8028 | 12.21 | 12.0062 |
| 24 | −1 | 1 | 0 | 60.5 | 57.7181 | 10.01 | 9.96042 |
| 25 | −1 | −1 | 0 | 45.7 | 43.9986 | 11.52 | 11.3385 |
| 26 | −1 | 1 | −1 | 54.7 | 52.9139 | 10.01 | 10.0894 |
| 27 | 0 | 1 | −1 | 56.6 | 52.4347 | 8.89 | 9.43708 |
| 28 | −1 | −1 | −1 | 36.8 | 39.5194 | 11.56 | 11.9699 |
| 29 | 0 | 1 | 1 | 61.8 | 65.2847 | 9.53 | 9.25042 |
| 30 | 0 | 1 | 0 | 59.7 | 57.1639 | 9.73 | 9.4075 |
| 31 | 0 | −1 | 0 | 44.3 | 42.3694 | 10.44 | 10.8047 |
| 32 | 1 | 1 | 0 | 54.3 | 54.0014 | 10.88 | 10.1571 |
| 33 | 0 | −1 | −1 | 33.2 | 37.9653 | 11.26 | 11.3368 |
| 34 | 1 | −1 | 0 | 37.4 | 38.1319 | 11.48 | 11.5735 |
| 35 | −1 | −1 | 1 | 50.2 | 51.8694 | 10.26 | 10.5795 |
| 36 | 1 | 1 | −1 | 50.1 | 49.3472 | 10.03 | 10.0873 |

## 4. Conclusions

In this research work, the immobilization of Lipozyme RM IM, a *Rhizomucor miehei* lipase, on different supports has been evaluated by its use as biocatalyst in the synthesis of Ecodiesel, a biodiesel-like biofuel obtained by selective transesterification of sunflower oil with ethanol. The best results have been obtained using a commercial silica gel 60 as support, achieving similar results to those obtained over the non-immobilized Lipozyme. Likewise, the optimization of the reaction parameters as well as the influence of these parameters in the enzymatic transesterification has been carried out by a statistical multifactorial design of experiments with three factors run by the Software StatGraphics. The analysis of variance showed that, in order to obtain an improvement in selectivity and kinematic viscosity, the reaction conditions should be an oil/ethanol molar ratio of 1/6 and a high amount of 10 N NaOH, 50 μL, whereas the amount of lipase employed depends on which parameter wanted to be optimized, i.e., 0.012 g of lipase to achieve the best selectivity value or 0.017 to achieve the best kinematic viscosity. Furthermore, the quadratic models obtained showed good results in terms of predicting the selectivity and viscosity of the investigated systems, as it was corroborated by experimental reactions.

The stability of the heterogeneous biocatalysts was studied by successive reactions in order to evaluate the feasibility and economic viability of its application in industrial scale. In this way, by means of simple physical immobilization of the Lipozyme on silica-gel, 15 reuses can be obtained without an evidence of activity loss. This fact allows us to assume that it is possible to perform a much greater number of reuses. These results are very remarkable, taking into account that the commercial Lipozyme RM IM, immobilized in an organic polymer is able to perform only 6 reuses before the total loss of activity.

Besides, an improvement in the operational separation of the biocatalyst from the reaction medium was obtained, since the commercial Lipozyme has to be separated by centrifugation, whereas the supported Lipozyme can be separated by decantation.

**Author Contributions:** This research article is part of the doctoral thesis of J.C., directed by D.L., and F.M.B., who in general conceived and designed the experiments. C.L., A.A.R., B.H. and R.E. made substantive intellectual contributions to this study, making substantial contributions to conception and design of it, as well as to the acquisition, analysis and interpretation of data. Furthermore, J.C., D.L. and R.E. wrote the paper. All the authors have been also involved in drafting and revising the manuscript, so that everyone has given final approval of the current version to be published in Energies Journal.

**Acknowledgments:** This research is supported by the MEIC funds (Project ENE 2016-81013-R), Junta de Andalucía and FEDER (P11-TEP-7723), that cover the costs to publish in open access.

**Conflicts of Interest:** The authors declare no conflict of interest.

## References

1. Ko, C.-H.; Chaiprapat, S.; Kim, L.-H.; Hadi, P.; Hsu, S.-C.; Leu, S.-Y. Carbon sequestration potential via energy harvesting from agricultural biomass residues in Mekong River basin, Southeast Asia. *Renew. Sustain. Energy Rev.* **2017**, *68*, 1051–1062. [CrossRef]
2. Joshi, G.; Pandey, J.K.; Rana, S.; Rawat, D.S. Challenges and opportunities for the application of biofuel. *Renew. Sustain. Energy Rev.* **2017**, *79*, 850–866. [CrossRef]
3. Sen, B.; Ercan, T.; Tatari, O. Does a battery-electric truck make a difference?—Life cycle emissions, costs, and externality analysis of alternative fuel-powered Class 8 heavy-duty trucks in the United States. *J. Clean. Prod.* **2017**, *141*, 110–121. [CrossRef]
4. Nair, S.; Paulose, H. Emergence of green business models: The case of algae biofuel for aviation. *Energy Policy* **2014**, *65*, 175–184. [CrossRef]
5. Noh, H.M.; Benito, A.; Alonso, G. Study of the current incentive rules and mechanisms to promote biofuel use in the EU and their possible application to the civil aviation sector. *Transp. Res. Part D Transp. Environ.* **2016**, *46*, 298–316. [CrossRef]
6. Winnes, H.; Styhre, L.; Fridell, E. Reducing GHG emissions from ships in port areas. *Res. Transp. Bus. Manag.* **2015**, *17*, 73–82. [CrossRef]
7. Russo, D.; Dassisti, M.; Lawlor, V.; Olabi, A. State of the art of biofuels from pure plant oil. *Renew. Sustain. Energy Rev.* **2012**, *16*, 4056–4070. [CrossRef]
8. Collet, P.; Lardon, L.; Hélias, A.; Bricout, S.; Lombaert-Valot, I.; Perrier, B.; Lépine, O.; Steyer, J.-P.; Bernard, O. Biodiesel from microalgae–Life cycle assessment and recommendations for potential improvements. *Renew. Energy* **2014**, *71*, 525–533. [CrossRef]
9. Dharma, S.; Masjuki, H.; Ong, H.C.; Sebayang, A.; Silitonga, A.; Kusumo, F.; Mahlia, T. Optimization of biodiesel production process for mixed Jatropha curcas–Ceiba pentandra biodiesel using response surface methodology. *Energy Convers. Manag.* **2016**, *115*, 178–190. [CrossRef]
10. Talebian-Kiakalaieh, A.; Amin, N.A.S.; Mazaheri, H. A review on novel processes of biodiesel production from waste cooking oil. *Appl. Energy* **2013**, *104*, 683–710. [CrossRef]
11. Wadumesthrige, K.; Ara, M.; Salley, S.O.; Ng, K.S. Investigation of lubricity characteristics of biodiesel in petroleum and synthetic fuel. *Energy Fuels* **2009**, *23*, 2229–2234. [CrossRef]
12. Aransiola, E.; Ojumu, T.; Oyekola, O.; Madzimbamuto, T.; Ikhu-Omoregbe, D. A review of current technology for biodiesel production: State of the art. *Biomass Bioenergy* **2014**, *61*, 276–297. [CrossRef]
13. Santori, G.; Di Nicola, G.; Moglie, M.; Polonara, F. A review analyzing the industrial biodiesel production practice starting from vegetable oil refining. *Appl. Energy* **2012**, *92*, 109–132. [CrossRef]

14. Quispe, C.A.; Coronado, C.J.; Carvalho, J.A., Jr. Glycerol: Production, consumption, prices, characterization and new trends in combustion. *Renew. Sustain. Energy Rev.* **2013**, *27*, 475–493. [CrossRef]
15. Ang, G.T.; Tan, K.T.; Lee, K.T. Recent development and economic analysis of glycerol-free processes via supercritical fluid transesterification for biodiesel production. *Renew. Sustain. Energy Rev.* **2014**, *31*, 61–70. [CrossRef]
16. Okoye, P.; Hameed, B. Review on recent progress in catalytic carboxylation and acetylation of glycerol as a byproduct of biodiesel production. *Renew. Sustain. Energy Rev.* **2016**, *53*, 558–574. [CrossRef]
17. Luna, D.; Calero, J.; Sancho, E.; Luna, C.; Posadillo, A.; Bautista, F.; Romero, A.; Berbel, J.; Verdugo, C. Technological challenges for the production of biodiesel in arid lands. *J. Arid Environ.* **2014**, *102*, 127–138. [CrossRef]
18. Calero, J.; Luna, D.; Sancho, E.D.; Luna, C.; Bautista, F.M.; Romero, A.A.; Posadillo, A.; Berbel, J.; Verdugo-Escamilla, C. An overview on glycerol-free processes for the production of renewable liquid biofuels, applicable in diesel engines. *Renew. Sustain. Energy Rev.* **2015**, *42*, 1437–1452. [CrossRef]
19. Ilham, Z.; Saka, S. Two-step supercritical dimethyl carbonate method for biodiesel production from Jatropha curcas oil. *Bioresour. Technol.* **2010**, *101*, 2735–2740. [CrossRef] [PubMed]
20. Casas, A.; Ruiz, J.R.; Ramos, M.A.J.S.; Pérez, A. Effects of triacetin on biodiesel quality. *Energy Fuels* **2010**, *24*, 4481–4489. [CrossRef]
21. Caballero, V.; Bautista, F.M.; Campelo, J.M.; Luna, D.; Marinas, J.M.; Romero, A.A.; Hidalgo, J.M.; Luque, R.; Macario, A.; Giordano, G. Sustainable preparation of a novel glycerol-free biofuel by using pig pancreatic lipase: Partial 1, 3-regiospecific alcoholysis of sunflower oil. *Process Biochem.* **2009**, *44*, 334–342. [CrossRef]
22. Luna, D.; Bautista, F.M.; Caballero, V.; Campelo, J.M.; Marinas, J.M.; Romero, A.A. Method for Producing Biodiesel Using Porcine Pancreatic Lipase as an Enzymatic Catalyst. WO Patent 2008009772A1, 24 January 2008.
23. Luna, C.; Luna, D.; Bautista, F.M.; Estevez, R.; Calero, J.; Posadillo, A.; Romero, A.A.; Sancho, E.D. Application of Enzymatic Extracts from a CALB Standard Strain as Biocatalyst within the Context of Conventional Biodiesel Production Optimization. *Molecules* **2017**, *22*, 2025. [CrossRef] [PubMed]
24. Luna, C.; Sancho, E.; Luna, D.; Caballero, V.; Calero, J.; Posadillo, A.; Verdugo, C.; Bautista, F.M.; Romero, A.A. Biofuel that keeps glycerol as monoglyceride by 1, 3-selective ethanolysis with pig pancreatic lipase covalently immobilized on AlPO4 support. *Energies* **2013**, *6*, 3879–3900. [CrossRef]
25. Verdugo, C.; Luna, D.; Posadillo, A.; Sancho, E.D.; Rodríguez, S.; Bautista, F.; Luque, R.; Marinas, J.M.; Romero, A.A. Production of a new second generation biodiesel with a low cost lipase derived from Thermomyces lanuginosus: Optimization by response surface methodology. *Catal. Today* **2011**, *167*, 107–112. [CrossRef]
26. Calero, J.; Verdugo, C.; Luna, D.; Sancho, E.D.; Luna, C.; Posadillo, A.; Bautista, F.M.; Romero, A.A. Selective ethanolysis of sunflower oil with Lipozyme RM IM, an immobilized Rhizomucor miehei lipase, to obtain a biodiesel-like biofuel, which avoids glycerol production through the monoglyceride formation. *New Biotechnol.* **2014**, *31*, 596–601. [CrossRef] [PubMed]
27. Luna, C.; Verdugo, C.; Sancho, E.D.; Luna, D.; Calero, J.; Posadillo, A.; Bautista, F.M.; Romero, A.A. A biofuel similar to biodiesel obtained by using a lipase from Rhizopus oryzae, optimized by response surface methodology. *Energies* **2014**, *7*, 3383–3399. [CrossRef]
28. Bautista, F.M.; Bravo, M.C.; Campelo, J.M.; García, A.; Luna, D.; Marinas, J.M.; Romero, A.A. Covalent immobilization of porcine pancreatic lipase on amorphous AlPO4 and other inorganic supports. *J. Chem. Technol. Biotechnol.* **1998**, *72*, 249–254. [CrossRef]
29. Verdugo, C.; Luque, R.; Luna, D.; Hidalgo, J.M.; Posadillo, A.; Sancho, E.D.; Rodriguez, S.; Ferreira-Dias, S.; Bautista, F.; Romero, A.A. A comprehensive study of reaction parameters in the enzymatic production of novel biofuels integrating glycerol into their composition. *Bioresour. Technol.* **2010**, *101*, 6657–6662. [CrossRef] [PubMed]
30. Tan, K.T.; Lee, K.T.; Mohamed, A.R. A glycerol-free process to produce biodiesel by supercritical methyl acetate technology: An optimization study via response surface methodology. *Bioresour. Technol.* **2010**, *101*, 965–969. [CrossRef] [PubMed]

*Article*

# Performance and Emission Quality Assessment in a Diesel Engine of Straight Castor and Sunflower Vegetable Oils, in Diesel/Gasoline/Oil Triple Blends

**Rafael Estevez [1], Laura Aguado-Deblas [1], Alejandro Posadillo [2], Beatriz Hurtado [1], Felipa M. Bautista [1], José M. Hidalgo [3], Carlos Luna [1], Juan Calero [1], Antonio A. Romero [1] and Diego Luna [1,*]**

1 Departamento de Química Orgánica, Universidad de Córdoba, Campus de Rabanales, Ed. Marie Curie, 14014 Córdoba, Spain; q72estor@uco.es (R.E.); aguadolaura8@gmail.com (L.A.-D.); b.hurtadocebrian@gmail.com (B.H.); qo1baruf@uco.es (F.M.B.); qo2luduc@uco.es (C.L.); p72camaj@gmail.com (J.C.); qo1rorea@uco.es (A.A.R.)

2 Seneca Green Catalyst S.L., Campus de Rabanales, 14014 Córdoba, Spain; seneca@uco.es

3 Unipetrol Centre for Research and Education, a.s., Unipetrol výzkumne vzdelávací centrum, a.s., Areál Chempark, Záluzí 1, 436 70 Litvínov, Czech Republic; jose.hidalgo@unicre.cz

* Correspondence: diego.luna@uco.es; Tel.: +34-957212065

Received: 21 May 2019; Accepted: 5 June 2019; Published: 7 June 2019

**Abstract:** This research evaluates the possibility of using straight oils such as castor oil, which is not suitable for food use, and sunflower oil, used as a standard reference for waste cooking oils, in blends with gasoline as second-generation biofuels. To this end, a study of the rheological properties of biofuels obtained from these double blends has been carried out. The aim is to take advantage of the different properties of gasoline, i.e., its low viscosity and its high energy density to obtain blends whose rheological properties allow the substitution of fossil diesel in high extent. The incorporation of fossil diesel to these gasoline/oil mixtures produces diesel/gasoline/oil triple blends, which exhibited the suitable rheological properties to be able to operate in conventional diesel engines. Therefore, the behavior of these blends has been evaluated in a conventional diesel engine, operating as an electricity generator. The triple blends allow the substitution of fossil diesel up to 40% with sunflower oil, and up to 25% with castor oil, with excellent power results achieved for blends in which diesel is substituted up to 40%, and also in fuel consumption at high demand in comparison to conventional fossil diesel. Besides, a significant reduction in the emission of pollutants has also been obtained with these triple blends.

**Keywords:** gasoline oil blends; castor oil; sunflower oil; biofuel; diesel engine; electricity generator; smoke opacity; Bacharach opacity; straight vegetable oils (SVO)

---

## 1. Introduction

Nowadays, fossil fuels constitute the main source of energy supply for transport. In fact, about 11 billion tons of fossil fuels are consumed each year worldwide [1]. Furthermore, it is expected that the fossil fuel demand will continue to rise, which will unavoidably lead to a scenario where fossil fuels run out [2]. In addition to the depletion of fossil fuels, as the demand for energy continues growing, the undesirable environmental effects linked to its production and consumption are becoming more evident by the day. In fact, emission of smoke, particulate matter (PM), carbon monoxide (CO), carbon dioxide ($CO_2$), nitrogen oxides ($NO_x$) and unburnt hydrocarbons (UBHC) from fossil fuel combustion are the primary causes of both atmospheric pollution and human health damage. For this reason, the generation of a safe, efficient and clean energy system is a priority objective [3]. Then, there is an urgent necessity for a transition from non-renewable and polluting energies, used up to now by society as a resource to guarantee their energy needs, to other renewable and environmentally sustainable alternatives [4]. However, in order to be competitive and viable, this energy transition model cannot ignore the actual vehicle fleet (more than one billion) operating with fossil fuels [5]. This is the reason why biofuels seem to be the right candidates to start the energetic transition abovementioned. The use of biofuels diminishes the fossil fuels depletion, minimizes the negative impact of Greenhouse Gases (GHG), and also allows using the current car fleet without any mechanic modification of compression-ignition (CI) or diesel engines [6]. In addition, biofuels can be easily integrated into the logistics of the global transport system, through the gradual replacement of fossil fuels by mixtures of diesel/biofuel. In this sense the EU stated that, in 2010, traffic fuels must contain at least 5.75% renewable bio-components, increasing this percentage up to 20% in 2020 and 30% in 2030. These measures foresee to achieve a reduction of 40% in the Greenhouse Gases emissions in comparison to those in the year 1990, with 27% of energy consumption from renewable sources and, at least, an increase of 27% in energy efficiency [7]. Despite these objectives that are apparently not difficult to achieve, replacement of fossil diesel with conventional biodiesel is still considered economically unfeasible, due to different factors associated to the biodiesel purification process, e.g., long reaction times and high energy consumption. Furthermore, during biodiesel production, glycerol is obtained as a by-product, being approximately 10 wt % of the total biodiesel produced.

To solve this problem, selective transesterification of triglycerides with ethanol has been described to produce monoglycerides (MG) as soluble glycerol derivatives using different lipases as catalysts [8]. Thus, through the partial transesterification of one mole of triglyceride (TG) with ethanol, two moles of ethyl esters (FAEE) and one mole of monoglyceride (MG) are generated, obtaining a biofuel called Ecodiesel that integrates glycerol in the form of a derivative soluble in the FAEE mixtures [8,9]. Recent studies stated that the presence of ethanol and other short-chain alcohols has a favorable effect on the emissions of the biofuels [10]. These mixtures improve the volatility of the fuel and constitute the so-called E-diesel, oxidiesel or oxygenated diesel, which in addition to reducing the emissions of the CI engines, improves the flow properties (viscosity) and the essential parameters that limit the application of diesel when operating at low temperatures, such as the "Cloud Point" (CP), "Pour Point" (PP), cold filter plugging point temperature (CFPP), point of occlusion of the cold filter (POFF), and emission levels of the motors without any significant negative effect in most of the parameters that define the quality of biodiesel [11].

In addition, the use of straight vegetable oils (SVO) in double blends with conventional diesel can be also considered as a potential option. All the relevant physicochemical properties of these blends are analogous to conventional diesel, except for the viscosity, which is much higher in oils than in diesel. Since fossil diesel has a much lower viscosity than oils, there will be a maximum percentage for each oil to be mixed with diesel in order to comply with regulations of the EN 14214 standard [12,13]. In this sense, it has been reported that blends with 10–20% of vegetable oil in diesel can be directly employed in diesel engines without any mechanic modification [14]. Following a strategy similar to that of E-diesel production, the incorporation of alcohols to form triple blends (diesel/biodiesel/alcohol) would further allow increasing the substitution of diesel. However, short-chain alcohols will have difficulty if to be blended with SVO due the different solubility, and also because a phase separation occurs after a short period of time, limiting the use of short-chain alcohols, mainly methanol and ethanol, with vegetable oils in the triple blends abovementioned. In addition, ethanol is corrosive and cannot be easily employed in today's engines or be shipped cheaply through current pipelines [15].

Nevertheless, there is a very interesting exception when castor oil is employed as an SVO due to the special structure of ricinoleic acid, which favors its solubility with alcohols, making possible a higher incorporation of them in triple blends. In fact, a diesel/castor oil/2-propanol triple blend in a proportion of 50/25/25 has been employed in conventional diesel engines, achieving very good results [16,17].

Regarding the use of vegetable oils, another strategy consisting the blending of them with less viscous and lower cetane (LVLC) has also been reported [18]. Pine oil (viscosity value of 1.3 cSt) has been employed in mixtures with castor oil to compensate for the high viscosity of it (226.2 cSt). The properties of castor oil and pine oil are mutually balanced causing a good balance of generated smokes [18], although the low cetane number of pine oil limits its amount in the blends (30% by volume) due to engine knocking problems. In addition to pine oil, eucalyptus, camphor and orange oils have been also considered as LVLCs [19]. Likewise, gasoline exhibits viscosity values sufficiently low to obtain important reductions in the viscosity of the oil/gasoline blends. Furthermore, the high energy density of gasoline, only slightly lower than diesel and of course higher than short chain alcohols, as well as its high availability, would allow its use as a blending agent in a more advantageous way than with other renewable compounds already described.

With this in mind, in this study, gasoline has been evaluated, for the first time, as a blending agent to produce a gasoline/oil mixture acting as a LVLC in blends with diesel. This is possible due to the high solubility of any type of vegetable oil with gasoline. The main goal of this research is to achieve a high substitution of fossil fuels, in a feasible way from a technical and economic point of view, and in a short period of time. To do so, two types of oils, which do not compete with food uses and which present high availability, have been chosen. On one hand, sunflower oil has been studied as a standard reference of waste cooking oils and, on the other hand, castor oil has been studied as a reference to oils which are not employed in food uses.

To obtain the optimum gasoline/oil mixture which can be blended with diesel, and maintaining the appropriate parameters of the EN 590 standard, the kinematic viscosity at 40 °C has been chosen as the most significant parameter, since this is the unique parameter that varies significantly with the proportions of the gasoline/oil blend. The effect of biofuel blends on the performance and emissions of internal combustion engines can be extremely complex to predict, because oils and gasoline show antithetic effects on engine performance in important parameters such as the cetane index (or energy density) and flash point, which promotes positive or negative interactions that are difficult to predict a priori. Furthermore, the cloud and pour point of the blends have been studied.

Once the adequate diesel/gasoline/oil blends were obtained, i.e., met the EN 590 standard parameters, they were tested in a conventional CI engine, operating as an electricity generator, as it is foreseeable that they exhibit different behaviors. The efficiency obtained is related to the effective electrical power, determined from the voltage and amperage generated by the engine. Furthermore, the contamination degree obtained from the opacity values of the generated smokes has been evaluated, as well as the fuel consumption of the different blends employed.

## 2. Materials and Methods

### 2.1. Double Blends of Gasoline/Oil and Triple Blends of Diesel/Gasoline/Oil

Commercial sunflower oil (food quality), locally obtained, and castor oil (Panreac, Castellar Del Valles, Spain) were blended with gasoline in a first step to obtain the double blends. The double blends which met the requirements of the EN 14214 standard for being employed as biofuels were blended with fossil diesel (from a Repsol service station) in different proportions to obtain the triple blends.

### 2.2. Characterization of the Biofuel Mixtures

The rheological properties that influence most in the correct performance of biofuels are the kinematic viscosity, measured at 40 °C, and cold flow properties. The cold flow properties are determined by several parameters that define its behavior at low temperatures, such as Cloud Point, Pour Point, and point of obstruction of the filter at low temperatures. At low temperatures, the formation of nuclei of solid crystals occurs, increasing in size as the temperatures decrease. The temperature at which the crystals become visible (diameter ≥ 0.5 mm) is defined as the Cloud Point, because the crystals typically form a cloudy cloud or suspension. The Cloud Point usually occurs at a temperature higher than the Pour Point. Solids and crystals grow quickly and block the passage of fuel lines and filters causing operational problems [20].

#### 2.2.1. Viscosity Measurements

The kinematic viscosity has been measured in an Ostwald–Cannon–Fenske capillary viscometer (Proton Routine Viscometer 33200, size 150), determining the time required for a certain volume of liquid to pass between two marked points on the instrument, placed in an upright position. From the flow time (t), expressed in seconds, we obtain the kinematic viscosity expressed in centistokes, $\upsilon = C \cdot t$, where C is the calibration constant of the measurement system, specified by the manufacturer (0.040350 $mm^2/s^2$ at 40 °C). All measures have been performed in duplicate and are presented as the average of both, with an experimental error less than 0.35%, as required by the standard ASTM (American Society for Testing and Materials) D2270-79 method for calculating the viscosity index from kinematic viscosity at 40 and 100 °C [16].

#### 2.2.2. Determination of the Pour Point and Cloud Point of Biofuels

The Pour Point and Cloud Point are determined by introducing the different double or triple blends, of different composition, in a glass tube having a flat bottom [16]. The tube was tightly closed with the help of a cork carrying a thermometer with a temperature measuring range of −36 to 120 °C. The tube was introduced in a digitally controlled temperature refrigerator for twenty-four hours. The tubes were brought out from time to time and checked until the oil did not show any movement when the jar was horizontally tilted for 5 s. After this time, the loss of transparency of the solution is evaluated. The appearance of turbidity in the samples is indicative that the Cloud Point temperature has been reached (Cloud Point). After a progressive decrease in temperature, the samples are kept under observation until they stop flowing (Pour Point).

#### 2.2.3. Energy Performance and Pollutant Emissions Generated in a Diesel Engine Electric Generator, Fueled with Different Biofuel Blends

Following a previously described experimental methodology [16], the mechanical and environmental characterization of a compression ignition diesel engine has been carried out, working at a rate of 3000 rpm coupled to an AYERBE electric generator, 5KVA, 230v type AY4000MN, for the generation of electricity, operating at a crankshaft constant rotation rate and under different degrees of demand for electrical power. This is achieved by connecting heating plates of 1000 watts each one (Figure 1a). This diesel engine will operate at a constant rate of rotation of the crankshaft and torque, so that the different values of electrical power obtained will be an exact consequence of the mechanical power obtained after the combustion of the corresponding biofuel. Different tests were obtained by providing to the engine double and triple mixtures of different biofuels. The electrical power generated can be easily determined from the product of the potential difference (or voltage) and the electric current intensity (or amperage), see Equation (1), both obtained by means of a voltmeter-ammeter.

$$\text{Electrical power generated (Watts)} = \text{voltage (Volts)} \times \text{amperage (Amps)}. \tag{1}$$

The consumption of the diesel engine was calculated estimating the speed of consumption of the biofuel studied with the engine working at a pre-determined demand of electric power. Thus, by operating under the same volume of fuel (0.5 L), different operation times are achieved. It must be indicated that the values represented are the average of at least three measurements. The error bars are not indicated in Figures for better display, although in none of the cases the error in the measurements exceeded the 5%.

The contamination degree is evaluated from the opacity of the smoke generated in the combustion process. This is obtained by using an opacimeter (TESTO 308, Arquitecsolar, Granada, Spain) under the operating conditions studied, Figure 1b [16]. The data here compiled are the media of three repeated measures, attaining an experimental error lower than 6%. The results obtained with the biofuels evaluated were compared with the measurements obtained when conventional diesel was fueled. The opacity value can be expressed as a percentage (being 100% totally cloudy and 0% totally clear) or as an equivalent number called the k value (Opacity Bacharach) in a scale which goes from white (0 Bacharach unit) to black (9 Bacharach units), as established by the ASTM D 2156-94, Standard Test Method for Smoke Density in Flue Gases from Burning Distillate Fuels [21].

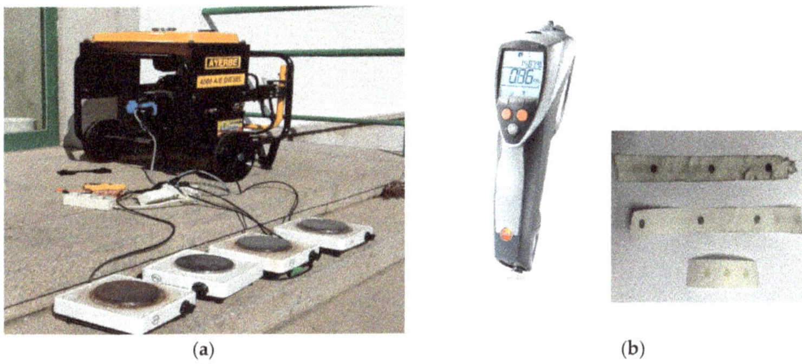


(a) (b)

**Figure 1.** Electrogenerator AYERBE, 5KVA, 230v tipo AY4000MN, heating plates of 1000 watts of power each and voltmeter-ammeter devise (yellow color, on the floor) (**a**); TESTO 308 opacity meter, which operates as established by the ASTM D 2156-94, Standard Test Method for Smoke Density in Flue Gases from Burning Distillate Fuels (**b**).

## 3. Results and Discussion

### 3.1. Rheological Properties of Gasoline/Oil Double Blends, and Diesel/Gasoline/Oil Triple Blends.

The major difference between diesel fuel and vegetable oils is the viscosity. In general, vegetable oils exhibit viscosity values in the range of 30–45 $mm^2/s$ or cSt, although castor oil has a much higher value of 227.0 cSt. For its part, fossil diesel exhibits values in the range of 2.5–6 cSt. The viscosity of the biofuel for being employed in a conventional diesel engine has to be in the range of 2.0 to 4.5 cSt (UNE EN ISO 3104). Therefore, viscosity is the essential parameter to modify in any vegetable oil, by mixing it with gasoline, for being employed in current diesel engines. In this respect, due to the importance of viscosity, its correct determination is critical to evaluate the quality of the biofuel. Thus, the viscosity, Cloud Point and Pour Point values of the different gasoline/oil blends studied are collected in Tables 1 and 2. As can be seen, an increase in the gasoline content in the blends promotes a decrease in their viscosity values, as well as a decrease in the Cloud and Pour Point of the double blends.

**Table 1.** Viscosity values at 40 °C (ASTM D2270-79), cloud point and pour point of the gasoline/sunflower oil blends. Errors are always calculated as the average of three measurements.

| Gasoline (%) | Viscosity (cSt) | Cloud Point (°C) | Pour Point (°C) |
|---|---|---|---|
| 0.0 | 37.80 | −7.0 ± 0.3 | −15.0 ± 0.6 |
| 10.0 | 20.64 | −8.0 ± 0.4 | −15.6 ± 0.6 |
| 15.0 | 17.07 | −8.6 ± 0.5 | −16.0 ± 0.6 |
| 20.0 | 13.01 | −9.1 ± 0.5 | −16.7 ± 0.7 |
| 25.0 | 9.67 | −10.0 ± 0.5 | −17.1 ± 0.6 |
| 30.0 | 7.52 | −10.5 ± 0.4 | −17.9 ± 0.7 |
| 35.0 | 5.73 | −11.2 ± 0.5 | −18.3 ± 0.7 |
| 40.0 | 4.42 | −12.0 ± 0.4 | −19.0 ± 0.7 |
| 45.0 | 3.46 | −12.8 ± 0.4 | −19.4 ± 0.8 |
| 50.0 | 2.74 | −13.6 ± 0.6 | −20.2 ± 0.8 |
| 100.0 | 1.17 | −19.0 ± 0.7 | −27.0 ± 1.1 |

**Table 2.** Viscosity at 40 °C (ASTM D2270-79), cloud point and pour point values of the gasoline/castor oil blends. Errors are always calculated as the average of three measurements.

| Gasoline (%) | Viscosity (cSt) | Cloud Point (°C) | Pour Point (°C) |
|---|---|---|---|
| 0.0 | 226.20 | −4.0 ± 0.2 | −24.0 ± 1.1 |
| 10.0 | 137.13 | −7.0 ± 0.3 | −24.2 ± 1.1 |
| 15.0 | 100.73 | −7.5 ± 0.3 | −24.3 ± 1.1 |
| 20.0 | 58.96 | −8.1 ± 0.4 | −24.7 ± 1 |
| 25.0 | 45.23 | −8.9 ± 0.4 | −24.9 ± 1.2 |
| 30.0 | 36.52 | −10.3 ± 0.5 | −25.0 ± 1.2 |
| 35.0 | 28.52 | −11.0 ± 0.5 | −25.2 ± 1.1 |
| 40.0 | 24.47 | −11.8 ± 0.6 | −25.4 ± 1.1 |
| 45.0 | 19.93 | −12.2 ± 0.6 | −25.7 ± 1.1 |
| 50.0 | 10.26 | −13.0 ± 0.6 | −25.8 ± 1.2 |
| 55.0 | 7.52 | −13.7 ± 0.7 | 25.9 ± 1.2 |
| 60.0 | 5.01 | −14.1 ± 0.7 | −26.0 ± 1.1 |
| 65.0 | 3.58 | −14.7 ± 0.8 | −26.1 ± 1.2 |
| 70.0 | 2.98 | −15.1 ± 0.8 | −26.2 ± 1.2 |
| 100.0 | 1.17 | −19.0 ± 1 | −27.0 ± 1.2 |

When sunflower oil is employed, optimum viscosity values for operating in compression ignition engines are achieved with the addition of 35–45% gasoline. In the case of castor oil, given its higher viscosity, it is necessary to incorporate a greater amount of gasoline in the blends (60–70%) to achieve an adequate viscosity value. Anyway, the gasoline/oil blends, in variable proportions, can be employed as a biofuel, even with oils which exhibit higher viscosity values, as is the case of castor oil. It must be highlighted that the presence of gasoline allows pure castor oil to be used without any chemical treatment, i.e., eliminating the transesterification to convert this oil in biodiesel and, therefore, considerably reducing the cost of the process. Be that as it may, the viscosity measurements indicate that it is possible to obtain biofuels with suitable rheological properties for being employed in diesel engines, complying with the European regulations EN 590, which stablish that viscosity at 40 °C must be in the range of 2.0–4.5 cSt.

Regarding the behavior of the double blends at low temperatures, the gasoline/oil blends began to solidify (Pour Point) at temperatures around −12 °C and they were completely frozen (Cloud Point) at −20 °C. This behavior is very similar to that exhibited by conventional fossil diesel, with temperatures around −10.0 °C and −18.5 °C for Pour Point and Cloud Point, respectively. However, the standard EN 590 classify the diesel fuel into two groups (with several subclasses) destined for specific climatic environments, depending on the geographical areas where it will be used. Therefore, the values of cloud point and pour point are in a wide range, between +5 and −34 °C.

As abovementioned, triple blends have been obtained by mixing different proportions of diesel with both a gasoline/sunflower oil blend with 40% of gasoline (Table 3) and also with a gasoline/castor oil blend with 60% of gasoline (Table 4). These triple blends are designated with the percentage that represents the corresponding gasoline/oil mixture, considering its biofuel character.

Hence, the B20 triple blend contains 80% of diesel, whereas the remaining 20% is a mix of gasoline and oil, in the proportions of the double blend previously studied. Therefore, with sunflower oil, since it is used in a double mixture containing 40% gasoline, the resulting triple blend corresponding to 80/8/12 diesel/gasoline/sunflower oil. In the case of castor oil, as the double mixture selected contains 60% of gasoline, the resulting triple blend corresponds to the proportion 80/12/8, diesel/gasoline/castor oil. The gasoline/oil double blends are represented as B100.

**Table 3.** Viscosity values of the triple blends, diesel/gasoline/sunflower oil, obtained by adding different amounts of fossil diesel to a double mixture of gasoline/sunflower oil, containing 40% gasoline.

| Nomenclature | Diesel/Gasoline/Sunflower Oil | Viscosity (cSt) |
|---|---|---|
| B20 | 80/8/12 | 3.94 |
| B30 | 70/12/18 | 4.06 |
| B40 | 60/16/24 | 4.06 |
| B50 | 50/20/30 | 4.18 |
| B60 | 40/24/36 | 4.29 |
| B70 | 30/28/42 | 4.41 |
| B80 | 20/32/48 | 4.53 |
| B100 | 0/40/60 | 4.42 |

**Table 4.** Viscosity values of the triple blends, diesel/gasoline/castor oil, obtained by adding different amounts of fossil diesel to a double mixture of gasoline/castor oil, containing 60% gasoline.

| Blends | Diesel/Gasoline/Castor Oil | Viscosity (cSt) |
|---|---|---|
| B20 | 80/12/8 | 4.49 |
| B30 | 70/18/12 | 3.82 |
| B40 | 60/24/16 | 3.82 |
| B50 | 50/30/20 | 3.70 |
| B60 | 40/36/24 | 3.70 |
| B70 | 30/42/28 | 3.46 |
| B80 | 20/48/32 | 3.22 |
| B100 | 0/60/40 | 5.01 |

### *3.2. Energy Performance and Pollutant Emissions Generated in a Diesel Engine Electric Generator, Fueled with Different Biofuel Blends*

As was expected, the triple blends of diesel/gasoline/sunflower oil and diesel/gasoline/castor oil exhibited adequate viscosity values to be employed as biofuels in a conventional diesel engine. Therefore, those blends were tested to ensure whether they present enough energy to guarantee the adequate performance of the engine. Furthermore, for comparative purposes, conventional diesel fuel was also employed as reference in the same conditions.

Thus, Figure 2 shows the power generated (Figure 2a) and the opacity in the smokes (Figure 2b) at different power demanded for the triple blends of diesel/gasoline/sunflower oil. In general, a stabilization of the power generated occurs between 3000 and 4000 W. In all the cases, at the highest value of power demanded (5000 W), a decrease in the power generated was observed. This generalized behavior could be explained by considering that the engine responds to the energy density of the fuel used, i.e., to the cetane index of the blend. Thus, the engine has a nominal capacity of 5000 W of electrical power, but this could only be achieved when the cetane number is 100, that is, when pure cetane is employed as fuel. In fact, if we apply the formula (generated power/power demanded) × 100 for the fossil diesel at the higher power-generated value (5000 W), we can estimate that its cetane number is around 52. This value is very close to the nominal value of the cetane number for fossil diesel (51), as it is collected in EN 590. Consequently, this could be a useful and simple procedure to calculate the cetane number of a fuel in a very approximate way.

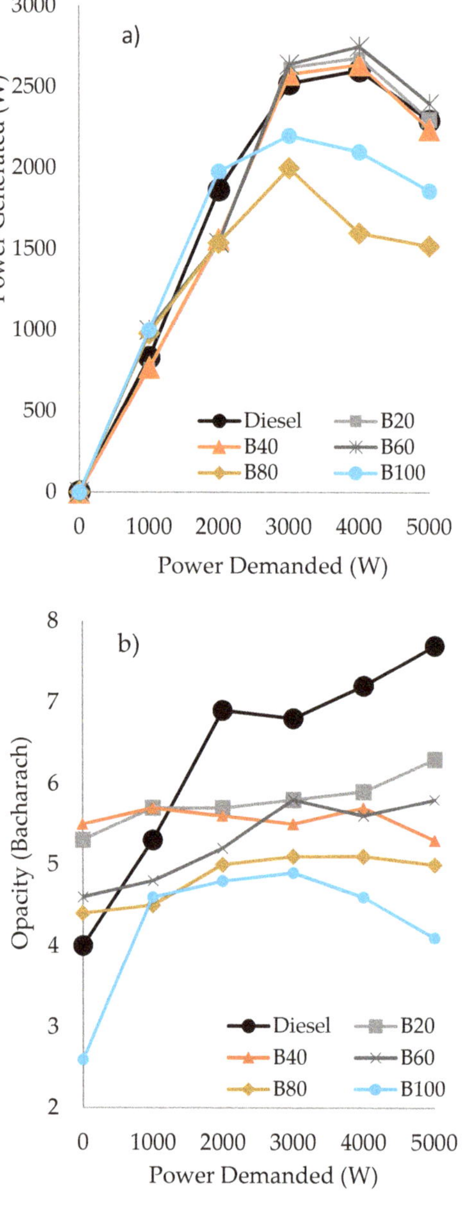


**Figure 2.** Power generated (in Watts), based on the power demanded (in Watts) (**a**) and opacity (in Bacharach) as a function of the power demanded (**b**) by the triple blends of diesel/gasoline/sunflower oil.

Regarding the behavior of the triple blends with sunflower oil (Figure 2a), the B20, B40 and B60 blends exhibited a similar performance than fossil diesel. However, the blends with lower or no amount of diesel in the blend, i.e., B80 and B100, generate lower power values than fossil diesel, which is in agreement with their lower cetane number. Therefore, it seems that, at least, a minimum amount of diesel (40%) is required for a good performance of the triple blends.

According to the contamination results obtained for the blends of diesel/gasoline/sunflower oil (Figure 2b), it can be seen that for all the triple blends, the opacity generated from 2000 W onwards was lower than with fossil diesel. In fact, the higher the amount of oil, the lower the opacity generated, achieving a considerably reduced opacity with the blend B100. It must be taken into account that all the units have been expressed in units according to ASTM D 2156-94, Standard Test Method for Smoke Density in Flue Gases from Burning Distillate Fuels. The reduction in soot opacities, considerably lower than diesel for the B80 and B100 triple blends, could be partly attributed to the power loss exhibited by them. However, for the blends B20, B40 and B60, these reductions can be mainly attributed to the biofuel chemical properties, as the power generated is similar to fossil diesel.

Considering the blends composed by diesel/gasoline/castor oil, similar results of power generated than those obtained with their counterparts using sunflower oil were obtained for the blends B80 and B100, see Figures 2a and 3a. However, the good results of power generated obtained with B40 and B60, even better than fossil diesel, has to be highlighted, whereas the B20 blend performed in a similar way than diesel. Regarding the opacity generated for the blends with castor oil, see Figure 3b, independently on the power demanded from 1000 W onward, all the blends performed better than fossil diesel. Furthermore, in comparison to the behavior observed with the blends of diesel/gasoline/sunflower oil, their counterparts using castor oil exhibited lower opacity values. Be that as it may, independently of the vegetable oil employed, a significant reduction in opacity values has been obtained, mainly at medium and high demand (from 2000 W onward), being a reduction in the range of 20–50% less than that obtained with diesel.

On the other hand, when a biofuel is considered for being employed in a diesel engine, another important factor to be taken into account is the consumption at different power demands. In this sense, the consumption of the different biofuels employed have also been evaluated and the results are plotted in Figure 4.

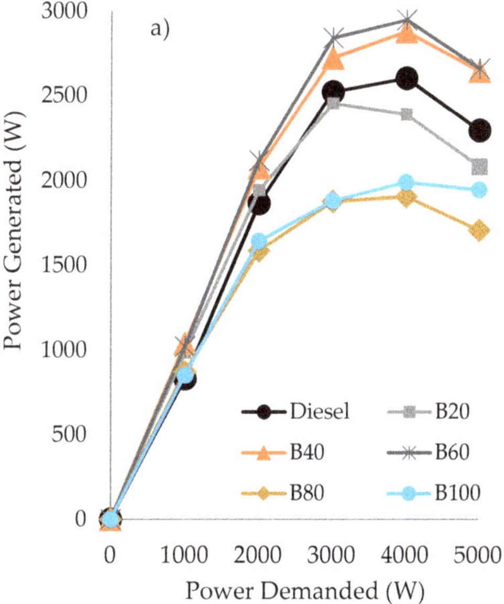


**Figure 3.** *Cont.*

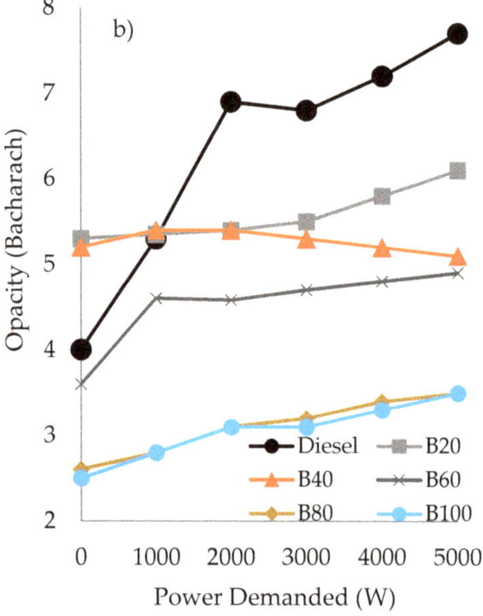


**Figure 3.** Power generated (in Watts), based on the power demanded (in Watts) (**a**) and opacity (in Bacharach) as a function of the power demanded (**b**) by the triple blends of diesel/gasoline/castor oil.

As can be seen, at low power demand (1000 W), the consumption of the blends is always higher than that obtained with fossil diesel, independently of the oil employed in the blend. However, at the highest power demanded (5000 W), the opposite behavior is observed. In fact, at 5000 W of power demanded, only the double blends of gasoline/oil exhibited higher consumption than diesel (20% higher with sunflower and 10% with castor oil). Anyway, it can be observed that the blends obtained using castor oil exhibited a lower consumption than their counterparts employing sunflower oil.

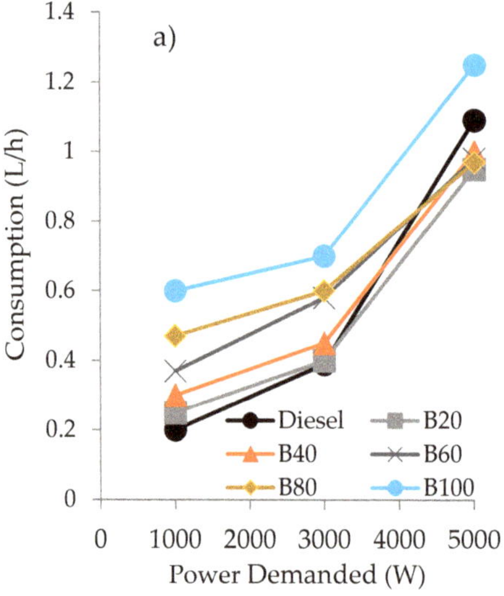


**Figure 4.** *Cont.*

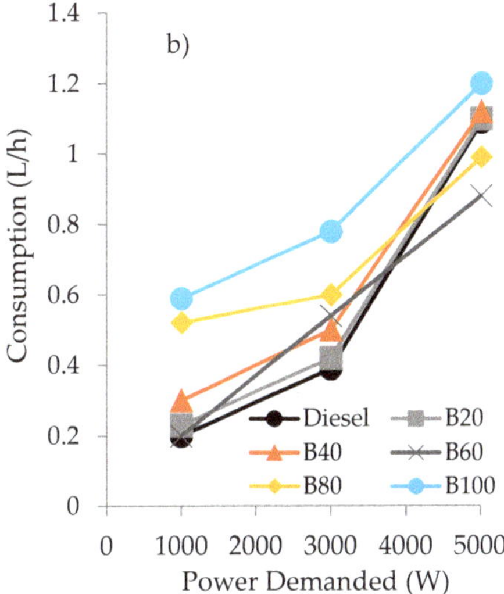


**Figure 4.** Consumption values as a function of the power demanded of the engine for the blends of diesel/gasoline/sunflower oil (**a**) and diesel/gasoline/castor oil (**b**).

## 4. Conclusions

In the present work, an alternative methodology to the chemical transformation of vegetable oil for being employed as biofuels is presented. This methodology consists of a dissolution process of the oil in a suitable solvent, in order to reduce its viscosity to the appropriate level demanded by the EN 590 standard. Specifically, we studied the use of gasoline as a solvent to obtain double blends of gasoline/oil, which are employed as LVLC in blends with fossil diesel. In this sense, two different commercial vegetable oils have been employed, sunflower oil and castor oil. According to the results here reported, the following conclusions have been obtained: The substitution of more than 40% of fossil diesel can be achieved by the use of triple blends of diesel/gasoline/sunflower oil, whereas 25% of fossil diesel can be substituted employing their counterparts with castor oil.

- With these blends, promising results of power generation, consumption and opacity of the smokes were obtained in comparison with those obtained with fossil diesel.
- Independently of the oil employed, the B20, B40 and B60 triple blends exhibited the best results in power generated. The lowest values of power generated obtained over B80 and B100 blends can be ascribed to the lower cetane number, as it is relative to the diesel in the blends.
- The opacity obtained with the different blends was always lower than that obtained with fossil diesel from a power demanded from 1000 W onwards.
- The consumption of the blends studied was higher than that obtained with diesel at lower and medium demand, i.e., 1000 and 3000 W, whereas at the highest demand here studied, 5000 W, the consumption is similar or even better with the diesel/gasoline blends, independently of the oil employed.

In summary, this research show, for the first time, that the substitution of diesel fuel for gasoline allows the incorporation of higher percentages of pure vegetable oils, generating second-generation biofuels capable of operating in current CI engines without making any mechanical modification in them. In addition, through the use of these triple blends of diesel/gasoline/vegetable oil, the objectives proposed by the EU in the next decades can be achieved, making possible the gradual substitution of fossil fuels by directly using vegetable oils, without performing any type of transesterification process. Therefore, this research opens a practical and economically viable alternative to the chemical production of biofuels.

**Author Contributions:** This research article is part of the doctoral thesis of B.H., directed by professors D.L. and F.M.B., who in a general way conceived and designed the experiments and wrote the paper. C.L., A.A.R., B.H.C. and R.E. made substantive intellectual contributions to this study, making substantial contributions to conception and design of it, as well as to the acquisition, analysis and interpretation of data. Furthermore, D.L. and R.E wrote the paper. All the authors have also been involved in drafting and revising the manuscript, so that everyone has given final approval of the current version to be published in the Energies journal.

**Funding:** This research received no external funding.

**Acknowledgments:** This research is supported by the MEIC funds (Project ENE 2016-81013-R), Junta de Andalucía and FEDER (P11-TEP-7723), that cover the costs to publish in open access.

**Conflicts of Interest:** The authors declare no conflict of interest.

## References

1. Covert, T.; Greenstone, M.; Knittel, C.R. Will we ever stop using fossil fuels? *J. Econ. Perspect.* **2016**, *30*, 117–138. [CrossRef]
2. Capellán-Pérez, I.; Mediavilla, M.; de Castro, C.; Carpintero, Ó.; Miguel, L.J. Fossil fuel depletion and socio-economic scenarios: An integrated approach. *Energy* **2014**, *77*, 641–666. [CrossRef]
3. Arutyunov, V.S.; Lisichkin, G.V. Energy resources of the 21st century: Problems and forecasts. Can renewable energy sources replace fossil fuels? *Russ. Chem. Rev.* **2017**, *86*, 777. [CrossRef]
4. Pang, X.; Mörtberg, U.; Brown, N. Energy models from a strategic environmental assessment perspective in an EU context—What is missing concerning renewables? *Renew. Sustain. Energy Rev.* **2014**, *33*, 353–362. [CrossRef]
5. Chu, S.; Majumdar, A. Opportunities and challenges for a sustainable energy future. *Nature* **2012**, *488*, 294. [CrossRef] [PubMed]
6. Lopes, M.; Serrano, L.; Ribeiro, I.; Cascão, P.; Pires, N.; Rafael, S.; Tarelho, L.; Monteiro, A.; Nunes, T.; Evtyugina, M. Emissions characterization from EURO 5 diesel/biodiesel passenger car operating under the new European driving cycle. *Atmos. Environ.* **2014**, *84*, 339–348. [CrossRef]
7. Dafnomilis, I.; Hoefnagels, R.; Pratama, Y.W.; Schott, D.L.; Lodewijks, G.; Junginger, M. Review of solid and liquid biofuel demand and supply in Northwest Europe towards 2030–A comparison of national and regional projections. *Renew. Sustain. Energy Rev.* **2017**, *78*, 31–45. [CrossRef]
8. Calero, J.; Luna, D.; Sancho, E.D.; Luna, C.; Bautista, F.M.; Romero, A.A.; Posadillo, A.; Berbel, J.; Verdugo-Escamilla, C. An overview on glycerol-free processes for the production of renewable liquid biofuels, applicable in diesel engines. *Renew. Sustain. Energy Rev.* **2015**, *42*, 1437–1452. [CrossRef]
9. Luna, D.; Calero, J.; Sancho, E.; Luna, C.; Posadillo, A.; Bautista, F.; Romero, A.; Berbel, J.; Verdugo, C. Technological challenges for the production of biodiesel in arid lands. *J. Arid Environ.* **2014**, *102*, 127–138. [CrossRef]
10. Sayin, C. Engine performance and exhaust gas emissions of methanol and ethanol–diesel blends. *Fuel* **2010**, *89*, 3410–3415. [CrossRef]
11. Aydın, F.; Öğüt, H. Effects of using ethanol-biodiesel-diesel fuel in single cylinder diesel engine to engine performance and emissions. *Renew. Energy* **2017**, *103*, 688–694. [CrossRef]
12. Rakopoulos, D.; Rakopoulos, C.; Giakoumis, E.; Dimaratos, A.; Founti, M. Comparative environmental behavior of bus engine operating on blends of diesel fuel with four straight vegetable oils of Greek origin: Sunflower, cottonseed, corn and olive. *Fuel* **2011**, *90*, 3439–3446. [CrossRef]
13. D'Alessandro, B.; Bidini, G.; Zampilli, M.; Laranci, P.; Bartocci, P.; Fantozzi, F. Straight and waste vegetable oil in engines: Review and experimental measurement of emissions, fuel consumption and injector fouling on a turbocharged commercial engine. *Fuel* **2016**, *182*, 198–209. [CrossRef]
14. Mat, S.C.; Idroas, M.; Hamid, M.; Zainal, Z. Performance and emissions of straight vegetable oils and its blends as a fuel in diesel engine: A review. *Renew. Sustain. Energy Rev.* **2018**, *82*, 808–823. [CrossRef]
15. Savage, N. The ideal biofuel: A biomass-based fuel needs to be cheap and energy dense. Gasoline sets a high standard. *Nature* **2011**, *474*, S9. [CrossRef] [PubMed]

16. Hurtado, B.; Posadillo, A.; Luna, D.; Bautista, F.; Hidalgo, J.; Luna, C.; Calero, J.; Romero, A.; Estevez, R. Synthesis, Performance and Emission Quality Assessment of Ecodiesel from Castor Oil in Diesel/Biofuel/Alcohol Triple Blends in a Diesel Engine. *Catalysts* **2019**, *9*, 40. [CrossRef]
17. Kania, D.; Yunus, R.; Omar, R.; Rashid, S.A.; Jan, B.M. A review of biolubricants in drilling fluids: Recent research, performance, and applications. *J. Pet. Sci. Eng.* **2015**, *135*, 177–184. [CrossRef]
18. Prakash, T.; Geo, V.E.; Martin, L.J.; Nagalingam, B. Evaluation of pine oil blending to improve the combustion of high viscous (castor oil) biofuel compared to castor oil biodiesel in a CI engine. *Heat Mass Transf.* **2019**, *55*, 1491–1501. [CrossRef]
19. Vallinayagam, R.; Vedharaj, S.; Yang, W.; Roberts, W.L.; Dibble, R.W. Feasibility of using less viscous and lower cetane (LVLC) fuels in a diesel engine: A review. *Renew. Sustain. Energy Rev.* **2015**, *51*, 1166–1190. [CrossRef]
20. Mohammadi, P.; Tabatabaei, M.; Nikbakht, A.M.; Esmaeili, Z. Improvement of the cold flow characteristics of biodiesel containing dissolved polymer wastes using acetone. *Biofuel Res. J.* **2014**, *1*, 26–29. [CrossRef]
21. Blanco, M.; Coello, J.; Maspoch, S.; Puigdomènech, A.; Peralta, X.; González, J.; Torres, J. Correlating Bacharach opacity in fuel oil exhaust. Prediction of the operating parameters that reduce it. *Oil Gas Sci. Technol.* **2000**, *55*, 533–541.

*Review*

# An Overview of the Production of Oxygenated Fuel Additives by Glycerol Etherification, Either with Isobutene or *tert*-Butyl Alcohol, over Heterogeneous Catalysts

**Rafael Estevez *, Laura Aguado-Deblas, Diego Luna and Felipa M. Bautista ***

Departamento de Química Orgánica, Ed. Marie Curie, Campus de Rabanales, Instituto Universitario de Nanoquímica (IUNAN), Universidad de Córdoba, 14071 Córdoba, Spain; aguadolaura8@gmail.com (L.A.-D.); qo1lumad@uco.es (D.L.)

* Correspondence: q72estor@uco.es (R.E.); qo1baruf@uco.es (F.M.B.); Tel.: +34-957-212-065 (F.M.B.)

Received: 21 May 2019; Accepted: 18 June 2019; Published: 19 June 2019

**Abstract:** Biodiesel production has considerably increased in recent decades, generating a surplus of crude glycerol, which is the main drawback for the economy of the process. To overcome this, many scientists have directed their efforts to transform glycerol, which has great potential as a platform molecule, into value-added products. A promising option is the preparation of oxygenate additives for fuel, in particular those obtained by the etherification reaction of glycerol with alcohols or olefins, mainly using heterogeneous catalysis. This review collects up-to-date research findings in the etherification of glycerol, either with isobutene (IB) or *tert*-Butyl alcohol (TBA), highlighting the best catalytic performances reported. Furthermore, the experimental sets employed for these reactions have been included in the present manuscript. Likewise, the characteristics of the glycerol ethers–(bio)fuel blends as well as their performances (e.g., quality of emissions, technical advantages or disadvantages, etc.) have been also compiled and discussed.

**Keywords:** glycerol; heterogeneous catalysis; etherification; isobutene; *tert*-Butyl alcohol; oxygenated fuel additives

## 1. Introduction

Nowadays, fossil fuels are the main energetic source in the world. Furthermore, as industrialization and modernization increase day by day, it is expected that the demand of these energetic resources will continue to increase. In fact, the data published by the Organization of the Petroleum Exporting Countries (OPEC), in January of 2017, stated that the world oil supply was 95.7 mbd, whereas in 2015 the average world oil supply was 92.7 mbd. The unavoidable decrease in the world crude oil reserves, together with the higher demand, will cause a progressive increase in price. Thus, the spot price of the OPEC basket (65.33 US$/b, according to the Market Indicators in November 2018) has supposed an increase of 33% in respect to the price established in November of 2016 [1]. Very recently, the U.S. Energy Information Administration (EIA) forecast the world crude oil prices to rise gradually, averaging $65 per barrel in 2020, whereas the world benchmark Brent crude oil will average $61 per barrel in 2019. In addition to those facts, some experts have also predicted that the current oil supply may be completely depleted by the year 2050 [2].

In addition to the economic aspect, environmentally friendly alternatives which are able to palliate and/or substitute the use of fossil fuels as the main energetic source must be taken into account. With this in mind, the EU stated that, in 2010, traffic fuels must contain at least 5.75% of renewable bio-components, increasing this percentage up to 20% in 2020 and 30% in 2030 [3]. Besides, the United States Department of Energy (DOE) has set among its objectives the replacement of 30% of fossil fuels with biofuels, as well as 25% of the industrial organic chemicals by biomass-derived chemical compound by 2025 [4].

Renewable energy sources such as solar, hydroelectric, wind, geothermal, and biomass can be considered as viable alternatives to conventional crude oil [5]. Among these alternatives, biomass has emerged as the most viable option to substitute fossil fuels for different reasons (e.g., it is a plentiful source of renewable carbon for the production of biofuels and fine chemicals) [6–9]. Furthermore, the energy production from biomass generates a lower amount of greenhouse gases than fossil fuels [10]. Therefore, the objective is the implementation of a biorefinery in which, as occurs in current refineries, biofuels, high-value chemicals, and bio-based products can be obtained from biomass (Figure 1).

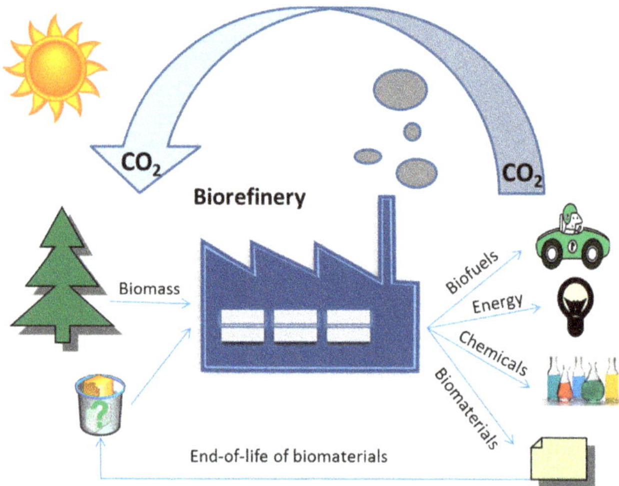


**Figure 1.** Diagram of a sustainable biorefinery process. Adapted from Ragauskas et al. [11] and Mohan et al. [12].

In the last two decades, the production of biofuels from biomass, especially to produce biodiesel, has received much attention. In fact, according to data reported by the EIA, the production of biodiesel in the United States increased in 2017 and 2018, Figure 2, despite the expiry of fiscal aids to the sector at the end of 2016 [13].

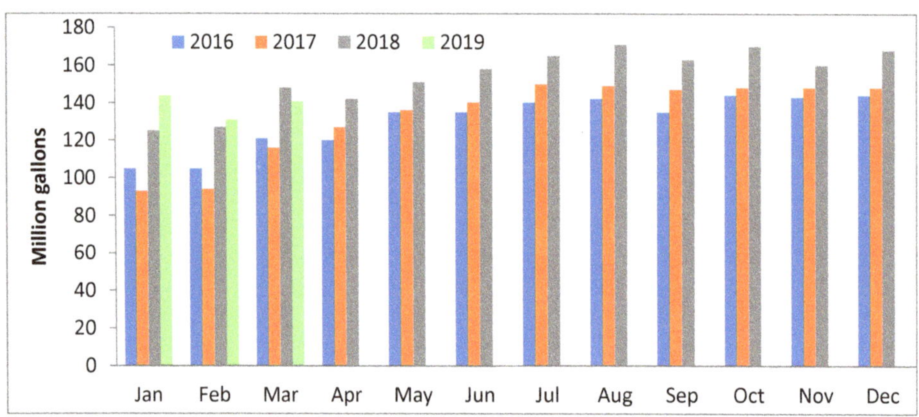


**Figure 2.** U.S. monthly biodiesel production 2016–2019 [13].

In fact, the increase in the production of biodiesel in the United States was decisive for the global biodiesel production (36.6 billion liters), around 1% higher than the previous year (34 billion liters) [14].

The skyrocketing production of biodiesel in recent decades is related to the inherent advantages that this biofuel exhibits to substitute fossil fuels, such as the low toxicity that it exhibits and its biodegradable, renewable, and biocompatible character. Furthermore, biodiesel can be easily integrated into the logistics of the global transportation system [15,16] and fits into existing engines with little or no modifications needed [17]. Biodiesel is obtained as a mixture of ethyl and methyl esters of fatty acids (FAEE or FAME), from vegetable oils or animal fats, through a transesterification reaction with a mono-alcohol, usually methanol, in the presence of an alkali homogeneous catalyst (Scheme 1). However, many researches are still being carried out, backed up by government policies, fiscal incentives, and emission laws, which aim to overcome different problems related to biodiesel production (e.g., economical and ethical aspects, as well as the valorization of the by-products generated). Hence, new catalytic systems and different feedstocks for biodiesel production are still being investigated [18–20].

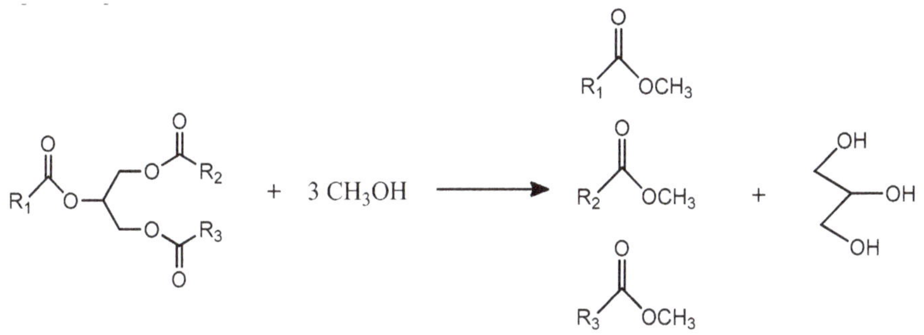

$$\text{triglyceride } (R_1, R_2, R_3) + 3\ CH_3OH \longrightarrow R_1COOCH_3 + R_2COOCH_3 + R_3COOCH_3 + \text{glycerol}$$

**Scheme 1.** Biodiesel production by transesterification of triglycerides with methanol.

Taking into account the biofuel synthesis and the available technology, the conventional biodiesel production, described by the standard EN 14214, presents the glycerol generated as a byproduct as the main drawback, which is 10% by weight of the biodiesel produced, constituting the major bottleneck in the production process [21]. Therefore, it is mandatory to obtain new and economically feasible ways of transforming glycerol, in order to enhance the sustainability of the biodiesel industries by increasing the crude glycerol value.

As it is well-known, glycerol exhibits a great versatility for being employed in different fields, such as cosmetic, food, pharmaceutical, and polymer industries and so on [2,22]. All these industries require glycerol with a high purity. However, crude glycerol from the biodiesel industry is a mixture of glycerol (~80%), water (~10%), NaCl (~10%), and methanol (<1%). Hence, a purification step of the crude glycerol phase to obtain a pure 88%–90% glycerol is mandatory. In addition to this, a further purification step is necessary to produce the pharmaceutical-grade glycerol meeting EU Pharmacopeia standard 99.5, thus making the process more expensive [23].

With this in mind, it is easy to understand the efforts carried out by the scientific community in order to obtain value-added products from glycerol, which can be introduced into the concept of biorefinery. In this sense, several recent reviews collected different approaches to transform glycerol into value-added products [2,21,24,25]. A summary of the applications of these glycerol derivatives is shown in Figure 3. In general, these products are obtained mainly by reactions in the presence of heterogeneous catalysts (e.g. hydrogenolysis, dehydration, esterification, etherification, acetalization, and so on) [26–28], making glycerol one of the most important platform molecules currently employed.

Among all the options to transform glycerol into value-added chemicals, the production of oxygenated fuel additives has gained greater attention in recent years [10,27–30]. By definition, a fuel additive is a chemical substance that can be blended with fuel (diesel, gasoline, and/or biodiesel) and it is capable of enhancing the engine performance by improving fuel properties, cleaning engine parts, reducing the consumption ratio, and/or decreasing greenhouse gas emissions [10,31].

Oxygenated fuel additives increase the octane rating and the combustion quality in the engine, since they reduce the particulate matter emission and the carbon monoxide production. The oxygenated fuel additives obtained from glycerol can be classified in three groups: Those obtained by glycerol acetylation, usually called acetin; the glycerol ethers; and the glycerol formal and ketals, obtained by the reaction of glycerol either with formaldehyde or acetone. These additives are considered as a good option to replace those petroleum-based additives, such as the methyl *tert*-Butyl ether (MTBE) and ethyl *tert*-Butyl ether (ETBE).

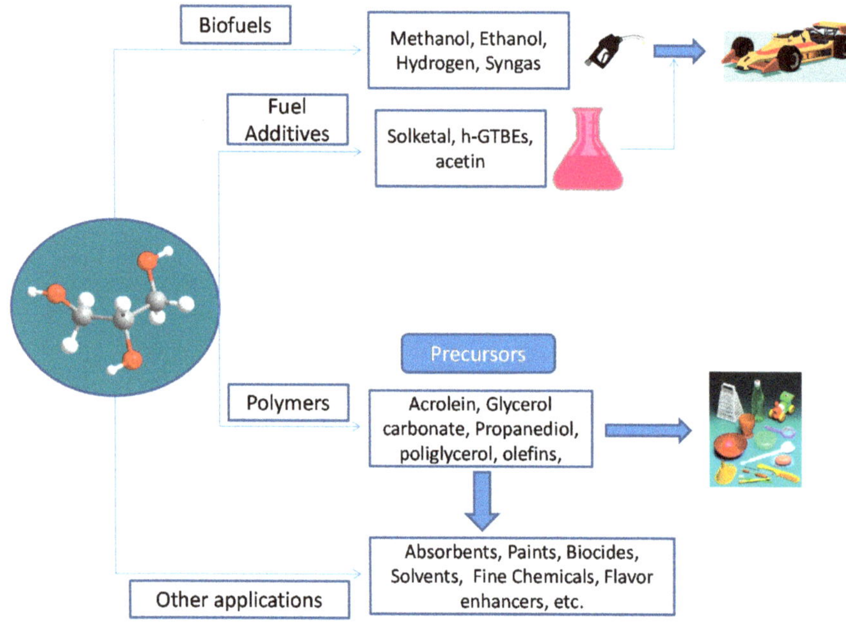


**Figure 3.** Main application fields of glycerol derivatives.

In fact, several researches have evaluated the effects of glycerol-derived oxygenated fuel additives on diesel engines [32–36]. One of the most promising fuel additives, taking into account the cost of production, is a mixture of di-*tert*-Butyl glycerol ethers (DTBGs) and tri-*tert*-Butyl glycerol ether (TTBG), the so-called h-GTBE, obtained by the etherification reaction of glycerol, either with isobutene (IB) or with *tert*-Butyl alcohol (TBA). This reaction has been studied mainly over heterogeneous catalysts due to the several advantages that they exhibit in comparison to homogeneous ones. For example, heterogeneous catalysts can be easily separated by a simple filtration or a centrifugation process. Then, the recovered catalyst can be employed again in subsequent reactions. Furthermore, the use of homogeneous catalysts can lead to different drawbacks such as the corrosion of the reaction systems (pipes, reactors, etc.), the necessity of a neutralization step if the use of mineral acids is required for the reaction to take place, and/or the impossibility of reusing the catalyst. Thus, despite the large number of applications of homogeneous catalysts, their substitution for heterogeneous ones with similar functionalities is becoming essential.

Thus, the present review mainly overviews the state-of-the-art production of glycerol *tert*-Butyl ethers either with IB or TBA on heterogeneous catalysts. Specifically, the influence exhibited by different catalytic processes and reaction conditions, as well as the characteristic of the catalysts on the values of conversion and yield to the desired products, have been compiled and discussed.

## 2. Etherification of Glycerol to Produce Oxygenated Fuel Additives

As is well-known, glycerol (G) cannot be added to fuels for different reasons. On the one hand, the high temperatures reached in the motor engines promote the glycerol polymerization as well as the formation of different products with high toxicity, such as acrolein. On the other hand, the low solubility of glycerol into fuels makes its use as fuel additive unsuitable. However, the glycerol ethers are able to be employed as oxygenated additives, as aforementioned. In addition, they can be used as industrial solvents and cleaning agents in industry [2].

### 2.1. *Etherification of Glycerol with Isobutene*

Scheme 2 shows the reaction pathways for the etherification of glycerol with isobutene (IB). It is accepted that the *tert*-Butylation of glycerol with isobutene needs of acid catalysis. The formation of the mono-*tert*-Butyl, di-*tert*-Butyl, and tri-*tert*-Butyl glycerol ethers occurs by consecutive reactions in which glycerol reacts with isobutene. Logically, the etherification of glycerol is preferred on primary hydroxyl groups, 1-mono-*tert*-Butyl glycerol ether (1-MTBG) > 2-mono-*tert*-Butyl glycerol ether (2-MTBG).

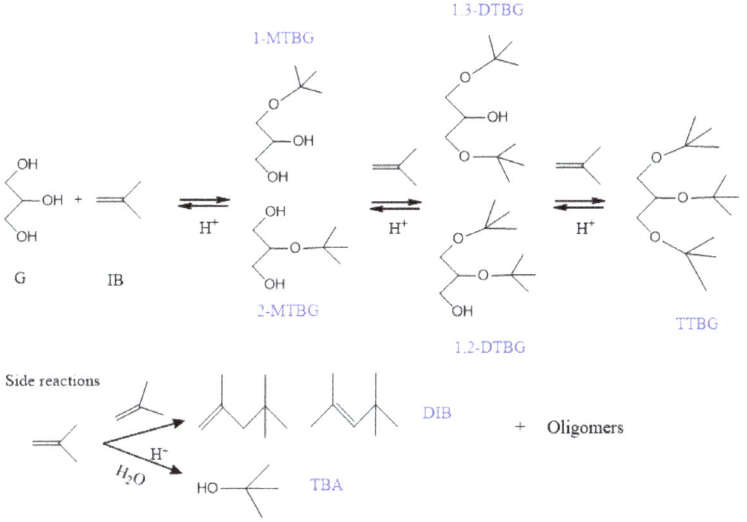

**Scheme 2.** Reaction pathways for the etherification of glycerol with isobutene.

In the beginning, this reaction was developed using p-toluenesulfonic acid and phosphorustungstic acid as homogeneous catalysts, obtaining high glycerol conversion values, 89% and 79%, respectively [37]. However, the heterogeneous catalysis is much more favorable considering operational, economic, and environmental aspects. Table 1 compiles the most relevant catalytic results to date in the glycerol etherification with IB. Thus, Karinen and Krause [38] studied the optimal reaction conditions for the etherification of glycerol with IB over a commercial ion-exchange macroreticular resin, the Amberlyst 35. They observed that the best results, regarding the formation of the DTBGs (60% of selectivity), were obtained at 80 °C reaction temperature and a stoichiometric molar ratio IB/G. A higher amount of initial G caused a mass transfer limitation owing to the high viscosity of the mixture, whereas an excess of IB favored its oligomerization. Furthermore, the authors observed that the addition of TBA to the reaction mixture improved the mass transfer between the phases, and also reduced the oligomerization of IB.

In the same line, Klepáčová et al. [39] reported the solventless etherification of glycerol with isobutene or *tert*-Butyl alcohol on different ion-exchange macroreticular resins, Amberlyst 15 and 35, and also over two large-pore zeolites, HY and H-Beta. Some important aspects were found in this research. On the one hand, the water obtained in the reaction of glycerol with *tert*-Butyl alcohol deactivated the acid sites of the ion exchange resins. On the other hand, if IB was employed as a reactant, an increase in the reaction temperature promoted a secondary reaction, such as the dimerization of isobutene, as well as the disproportion of the glycerol ethers formed. Regarding the behavior of the catalysts, the best results were obtained over the acid macroreticular ion exchange resins (100% of glycerol conversion and 92% of selectivity to h-GTBE) using IB as a reactant at the following reaction conditions: 60 °C, 7.5 wt.% of catalyst, 8 h, and a IB/G molar ratio of 4:1. In addition, the authors concluded that zeolites are not suitable catalysts for this reaction because a rapid deactivation occurred due to the small size of their pores, although good results of yield to DTBGs were obtained, 65.8% and 80.6% for HY and H-Beta, respectively. The same catalysts were tested employing a solvent, observing a change in the polarity and an homogenization of the reaction mixture [40]. As a consequence, the best results were obtained over on the HY zeolite (90% and 0.18 mol·$L^{-1}$ of DTBGs) after 8 h of reaction. However, in the presence of a solvent, TTBG was never obtained, corroborating the influence of the pore size. According to the author, and from a technological point of view, zeolites would not be suitable catalysts for this reaction due to the fact that they are easily deactivated and also due to the high prices that they exhibit in comparison to the ion exchange resins.

The possibility of partially neutralizing the acidity of A-15, in order to suppress the isobutene oligomerization, was studied by Lee et al. [41,42]. They observed that a partial substitution of acidic protons with $Na^+$, $Ag^+$, $Mg^{2+}$, and $Al^{3+}$ gave rise to a decrease in the rate of all the processes involved in the reaction, including the IB oligomerization that was efficiently suppressed, obtaining less than 50% of di-isobutene (DIB). Very recently, Bozkurt et al. [43] studied the influence of the degree of sodium-exchange in Amberlyst-15 acidity, evaluating the behavior of these catalysts in batch and flow reactors. Authors observed from the ammonium-temperature programmed desorption (TPD) measurements and density functional dispersion (DFT) calculations, that a substitution of the protons of the sulfonic groups in A-15 with $Na^+$ cations promoted a decrease in acidity, as was expected, but also promoted an increase in the strength of these acid sites. Thus, in batch reactors, a total conversion was attained, the higher the Na-exchange in A-15, the higher the selectivity to h-GTBE. Furthermore, the IB oligomerization was suppressed in almost one order of magnitude for the catalyst with the highest Na exchange. In addition, to elucidate the changes in the reaction network, the authors performed measurements in a once-through flow reactor at low conversions. The results showed that mono-*tert*-Butyl glycerol ethers (MTBGs) and DIB were the first products obtained on A-15, while DTBGs and TTBG also became primary products on sodium-exchange counterpart catalysts.

Melero et al. [44] studied the reaction over different sulfonic acid-functionalized mesostructured silicas, attaining higher yields than those obtained over commercial ion-exchange resins (A-15 and A-36). The optimal experimental conditions were chosen by statistical multivariate analysis, obtaining 92% of yield to h-GTBE over an arylsulfonic-SBA-15 at 75 °C, a molar ratio IB/G of 4, and 4 h of reaction. The absence of IB oligomers, despite an excess of IB (molar ratio IB/G = 4) being employed, contradicts the previous results reported by Karinen and Krause on the A-15 [38]. The authors explained this behavior was due to the moderate strength of the acid sites, as well as their location on the mesostructured silicas.

Likewise, several modified zeolites have been employed as catalysts in this reaction. Zhao et al. [45] performed the reaction over rare earth ($La^{3+}$, $Ce^{3+}$, $Nd^{3+}$, and $Eu^{3+}$) modified Hβ zeolites prepared by ion exchange. Among all the modified Hβ zeolites, the best catalytic results, 67% of yield to h-GTBE, were obtained over that modified with Neodymium at 70 °C; IB/G molar ratio of 3 and 2 h of reaction time. The highest activity attained on this catalyst is related to the highest acidity that it also exhibited. Xiao et al. [46] studied the modification of an HY zeolite by acid washing with citric acid and nitric acid, observing an improvement in the catalytic activity in comparison with the unmodified HY zeolite, that was associated with the modification of its textural properties, although the influence of the changes in acidity also observed should not be ruled out. González et al. [47] proposed the functionalization with sulfonic groups of several zeolites, such as Beta, ZSM-5, mordenite, and a commercial montmorillonite K-10. The incorporation of sulfonic acid groups was different on each zeolite, mainly due to the dealumination degree suffered by the zeolites as a consequence of the acid treatment. Be that as it may, the presence of sulfonic acid sites improved the catalytic activity of the zeolites studied. The influence of the microwave irradiation in the sulfonation of a Beta zeolite and also of a SBA-15 was also studied [48]. Independently of the heating method (e.g., microwave irradiation or conventional heating), complete conversion and a high selectivity to TTBG ethers (32-36%) were obtained on the functionalized beta zeolites. Hence, those results demonstrated that the microporosity of the zeolite was not difficult in the formation of TTBG if the material exhibited an adequate amount and strength of acid sites. Mesoporous silicas, MCM-41, SBA-15, and HMS, modified by different methodologies (e.g., introducing aluminum into the structure, incorporating phosphorous species or sulfonic groups) was also studied by the same authors [49]. The best results (100% of glycerol conversion and 84% of selectivity to h-GTBE) were obtained over the sulfonated HMS silica synthesized using dodecylamine (dda) as a surfactant. They concluded that, despite the fact that the aluminum and phosphorous incorporation improved the acidity of the catalysts, this was not strong enough to achieve a satisfactory selectivity to h-GTBE (<50%). Furthermore, the catalysts suffered some deactivation due to the reagents and reaction products blocked the pores. In order to avoid the deactivation suffered by these micro and mesoporous materials, González et al. [50] dealt with the synthesis of sulfonic acid-functionalized aerogels in a subsequent research. The best catalytic results (100% conversion and 75% selectivity to h-GTBE) were achieved over a sulfonated aerogel (MwS-AG) at 75 °C of temperature, an IB/G molar ratio of 4 and 24 h of reaction time. In addition, the catalyst was stable after four cycles.

Frusteri et al. [51–53] studied the catalytic performance of Hyflon®perfluorosulfonic ionomers supported on several spherical silicas in the etherification of glycerol with IB, attaining a yield to h-GTBE over Hyflon®/ES70Y of 89%. According to the authors, factors such as a strong ionomer–silica interaction, a high accessibility to the active centers of the catalysts, and also hydrophobic surface properties are the influential factors on the catalytic behavior. In order to further shift the equilibrium of the reaction the formation of h-GTBE, a new charge of catalyst was added to the final reaction mixture, with a decrease taking place in the amount of MTBGs in the mixture from 28.3% to 9.9% [52].

Catalysts based on carbon have also been reported in this reaction. Zhao et al. [54] proposed the synthesis of an amorphous carbon-based catalyst obtained by the sulfonation of partially carbonized peanut shells. This catalyst showed a very good catalytic performance in the reaction, achieving total conversion of glycerol and selectivity to h-GTBE of 92% at 70 °C, an IB/G molar ratio of 4, and 2 h of reaction time. Zhou et al. [55] synthesized a sulfonated graphene catalyst by grafting aryl radicals containing sulfonic groups onto the two-dimensional surface of graphene. This catalyst exhibited an almost total conversion of glycerol and a selectivity value to TTBG of 59%, at 70 °C, an IB/G molar ratio of 6, and 7 h of reaction time. Furthermore, these authors reported a simple methodology to obtain the desired ethers from the reaction mixture, consisting in the extraction of the non-desired products using fresh glycerol.

Voicu et al. [56] evaluated the effect of different types of emulsifiers in the reaction medium, employing a hydrated silicotungstic acid as the catalysts. Of all the emulsifiers employed, the best results were obtained using an amphoteric ammonium quaternary salt ($C_{19}$). This amphoteric emulsifier improved both the glycerol conversion and the selectivity to di- and tri-*tert*-Butyl glycerol ethers, reducing the dimerization of IB.

**Table 1.** Compilation of the best heterogeneous catalytic results on the glycerol etherification with IB.

| Catalyst [a] | Reaction Conditions | | | | | $X_G$ (mol %) | $S_{h\text{-}GTBE}$ (mol %) | Ref. |
|---|---|---|---|---|---|---|---|---|
| | T (°C) | P [b] (bar) | Cat. Loading (wt.% of G) | IB/G Mol/mol | Time (h) | | | |
| p-toluenesulfonic acid | 90 | 1.4 | 2.16 | 2 | 5 | 89 | 47 | [37] |
| Amberlyst-15 (A-15) | 80 | 15 | 1 g | 4 | 7 | >95 | 97 | [38] |
| Amberlyst-39 wet | 60 | Autoge | 7.5 | 4 | 8 | 100 | 93 | [39] |
| A-15 in dioxane | 60 | Autoge | 7.5 | 4 | 8 | 79 | 47 | [40] |
| Ag(62)A-15 [c], powder | 60 | 20 | 7.5 | 4 | 20 | >90 | 92 | [41] |
| Na(51)A-15 [c] | 60 | 20 | 7.5 | 4 | 20 | 99 | 90 | [42] |
| 0.3MNa-exchange A-15 | 75 | 10–15 | 7.5 | 3 | 6 | 100 | 92 | [43] |
| Ar-SBA-15 | 75 | 8 + VLE | 5 | 4 | 4 | 100 | 92 | [44] |
| Zeolite β + $Nd^{+3}$ | 70 | 15 | 6 | 3 | 2 | 93 | 75 | [45] |
| Zeolite Y+1M citric acid [c] | 80 | Autoge | 1 | 4 | 5 | 82 | 57 | [46] |
| Zeolite β-MwS(1.4) | 75 | 10 + VLE | 0.5 g | 4 | 48 | 100 | 90 | [47,48] |
| HMS(dda)-S | 75 | 10 + VLE | 0.5 g | 4 | 24 | 100 | 84 | [49] |
| MwS-AG | 75 | 10 + VLE | 0.5 g | 4 | 24 | 99 | 75 | [50] |
| 730SS1 | 70 | Autoge | 7.5 | 3 | 17 | 100 | 89 | [51] |
| Hyflon®/ES70Y | 70 | Autoge | 7.5 | 3 | 17 | 100 | 93 | [52] |
| Sulfonated peanut shell | 70 | 15 | 6 | 4 | 2 | 100 | 92 | [54] |
| SG | 70 | 10 | 2 | 6 | 7 | 99 | 96 | [55] |
| HSiW·$20H_2O$, in 0.7% $C_{19}$ | 80 | Auto | 5 | 3 | 5 | 99 | 90 | [56] |

a. Ar-SBA-15: Arenesulfonic-SBA; HMS (dda): HMS synthesized using dodecylamine as surfactant; MwS-AG: Microwave sulfonated silica aerogel; 730SS1: 730 equivalent weights on spherical silica, obtained by incipient wetness; Hyflon®/ES70Y: Hyflon Ion S4X perfluorosulfonic ionomer supported on spherical silica (ES70Y); SG: Sulfonated graphene; HSiW·$20H_2O$, in 0.7% $C_{19}$: silicotungstic acid hydrate in amphoteric emulsifier ($C_{19}$). b. Autoge: Autogenous pressure; VLE: Pressurizing the reactor and then to reach the vapor liquid equilibrium. c. Dioxane was employed as solvent.

Apart from the type of catalyst employed, some studies have focused their attention on the mass transfer and on the phase separation of the reaction mixture. This fact is of great importance owing to glycerol being immiscible in IB and in high ethers, as reported by Behr and Obendorf [37]. Thus, the biphasic system can be changed to a monophasic system by decreasing glycerol and increasing MTBG/h-GTBE or decreasing both glycerol and IB, while increasing mono-*tert*-Butyl glycerol ethers. In fact, they observed that after 40 min, the biphasic system became monophasic with a reaction mixture composition of 40 wt.% DTBG, 30 wt.% MTBG, and 5 wt.% of glycerol (IB and TTBG were not observed). This prototype was improved by Liu et al. [57,58], demonstrating that the *tert*-Butyl ethers of glycerol are soluble with each other in the temperature range of 70.2–100.2 °C. In addition, MTBG and glycerol are perfectly mixed under all concentrations tested, whereas the h-GTBE were partially soluble with glycerol, rising slightly when the temperature increased.

Experimental Designs Reported to Date for Glycerol Etherification with IB

Regarding the process designs and experimental settings, very few studies on the production of glycerol ether in continuous reactors, over heterogeneous catalysts, have been published to date [43]. Di Serio et al. [59] have reported a process for the production of h-GTBE employing the Amberlist-15 as a catalyst, in which those desired products were extracted by using biodiesel. This blend of biodiesel/h-GTBE can be employed directly as a diesel additive. More recently, Liu et al. [60] obtained the h-GTBE employing an ion exchange resin, NKC-9, as catalyst. After the reaction, the MTBGs were extracted with fresh glycerol, whereas residual glycerol of the reaction was washed with water. Furthermore, the IB and DIB produced during the reaction were separated by distillation, whereas the residual glycerol, after washing with water, and the recovered IB were recirculated in order to be employed again as reactants. Furthermore, the authors carried out a study on the feasibility of the process using a mathematical model, which included the cost related to the raw materials, the equipment, utilities, operating, and maintenance, concluding that the production of these additives can be very promising from the economical point of view if the IB price is below 1.2 $/Kg, considering that the cost to produce the glycerol ethers is 1 $/Kg. Regarding the possibility of using crude glycerol and IB from the $C_4$ fraction of crude oil, very few studies have been published [61,62], and only at laboratory scale.

### 2.2. Etherification of Glycerol with tert-Butyl Alcohol

In recent years, the possibility of using TBA instead of isobutene to obtain the glycerol *tert*-Butyl ethers has gained importance. There are different reasons to explain this fact. Firstly, isobutene is obtained in the catalytic cracking of crude oils. Concretely, 1-butene, cis/trans-2-butene, and IB conform the crude oil $C_4$ fraction. Therefore, their price depends directly on the crude prices, in addition to coming from a non-renewable source. Secondly, in order to keep the IB in the liquid phase in the etherification reaction, the use of additional pressure is required. In fact, as aforementioned, the mass transfer between the two phases is a limiting factor in this reaction, making the use of a solvent advisable to achieve a better catalytic performance. Regarding environmental aspects, the solventless reaction has to be promoted. Last but not least, an important drawback is the secondary reactions of IB, promoted by strong acid sites, giving rise to a decrease in the selectivity to the desired products, as aforementioned.

All these disadvantages can be palliated by using *tert*-Butyl alcohol, which is liquid at the reaction temperatures usually employed, allowing the mass transfer between the phases, acting itself as both solvent and reactant, avoiding the use of solvents capable of dissolving glycerol, with the technological problems that this fact implies, as well as typical drawbacks of a complex three-phase system. Likewise, the oligomerization of IB can be diminished, and the use of high pressure to carry out the reaction is not needed. Furthermore, TBA is currently obtained from the polypropylene production.

Analogously to the etherification of glycerol with IB, the etherification with TBA requires catalysts with acid sites, see Scheme 3. It is generally accepted that the reaction mechanism occurs via a rapid protonation on acid sites of TBA, which, on losing a water molecule, generates a tertiary carbocation. This carbocation reacts with glycerol, generating the MTBGs in the first step of the reaction. Analogously, the reaction of another TBA molecule with mono-*tert*-Butyl glycerol ethers generates the di-substituted ones (DTBGs), which, reacting with TBA, produces the TTBG in a subsequent reaction. Hence, water is formed in every step of the reaction, and IB can be also obtained by TBA dehydration. Hence, there are two aspects that must be taken into account employing TBA as a reactant. Firstly, the reaction is controlled by the thermodynamic equilibrium, which limits the maximum yield to the reaction products. Secondly, the formation of water in every step, Scheme 3, can promote the hydrolysis of the ether bonds formed, shifting the equilibrium to the left. In addition, the water molecules can solvate the acid sites, promoting a faster deactivation of the catalysts.

Thus, Klepàcová et al. [39] studied the etherification reaction with IB and TBA over two different ion exchange resins, A-15 and A-35, at two temperatures, 60 and 90 °C. The highest yield to h-GTBE (88.7%) was obtained over A-35 with IB, whereas the yield value using TBA was only 6.5%. Similar results were reported by González et al. [63] over several zeolitic materials. Klepàcová et al. [64] also reported the optimal experimental conditions for the reaction (i.e., 75 °C of reaction temperature and a high TBA/G molar ratio), which is logical according to the Le Chatelier principle. In fact, as can be seen in Table 2, temperatures in the range of 60–120 °C and a TBA/G molar ratio of 4 are the most typical conditions employed by the researchers. However, Chang et al. [65] reported a one-step synthesis and separation of the h-GTBE, consisting in employing very high temperatures (~190 °C), obtaining the reaction products separated in two phases. From the organic phase, the h-GTBEs were separated from IB and its oligomers by distillation below 225 °C, directly obtaining the oxygenated additives.

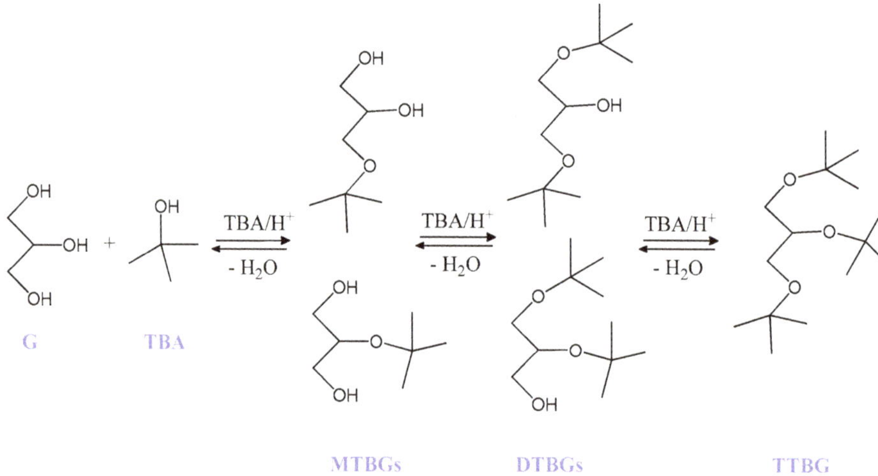

G TBA TBA/H+ - $H_2O$ TBA/H+ - $H_2O$ TBA/H+ - $H_2O$ MTBGs DTBGs TTBG

**Scheme 3.** Reaction pathways for the glycerol etherification with TBA.

In addition to the high TBA/G molar ratio, another strategy employed to shift the equilibrium to the formation of the desired products is the elimination of the water formed. Thus, Frusteri et al. [66] performed the reaction over silica-based catalysts, taking two ion exchange resins, SAC-13 and A-15, as reference. The best results were obtained over A-15 (yield to h-GTBE = 28%). The authors proposed two main reasons for the low formation of TTBG: the steric hindrance and the water formation during the reaction. In order to corroborate this, they proposed an experiment consisting of stopping the reaction after 6 h, to dehydrate the reaction mixture by adding zeolites, and then running again for another 6 h, obtaining an increase in DTBGs from 28% to 41%. This result clearly demonstrated that the presence of water makes the formation of h-GTBE difficult, in agreement with Klepàcová et al. [64]. To overcome the problem of the presence of water in the reaction mixture, Cannilla et al. [67,68] employed a water permselective membrane, allowing the equilibrium to be shifted toward the formation of poly-ethers. This membrane selectively removed the water formed by the recirculation of the gas phase. Acid catalysts, prepared by the impregnation of an ethanolic solution of Hyflon Ion S4X perfluorosulfonic ionomer on microspherical silica, were compared with Amberlist-15 [67], obtaining a better catalytic performance on the Hyflon supported catalysts with the use of the membrane, because of its higher hydrophilicity and a better accessibility to the active sites. Furthermore, the use of a membrane able to remove water selectively allowed thermodynamics constrains to be overcome, attaining total glycerol conversion and selectivity values to h-GTBE of 70%.

Apart from the ion exchange resins, other catalytic systems have been studied in order to overcome the limitations of these types of resins, such as the lack of thermal stability, the fact that they can be swelled and shrunk in organic media, and their hydrophobic character, since the sulfonic acid groups are the only hydrophilic part of the structure. The hydrophobic character promotes a faster deactivation by the generation of water, since this water solvates the sulfonic acid groups [66]. Thus, catalytic systems based on carbon, mainly prepared from biomass and sulfonated by different strategies, have gained importance. Thus, Gonçalves et al. [69] evaluated the behavior of sulfonated carbons prepared from agroindustrial wastes, showing a high catalytic performance. In fact, with the sugar cane bagasse-based catalyst, a glycerol conversion value of 80.9% and selectivity to h-GTBE of 21.3% were attained after 4 h of reaction time, at 120 °C, and using a 5 wt.% of catalyst loading. The hydrophilic character of the carbons makes possible the absorption of the water generated during the reaction, palliating the negative influence that this water usually has in the catalytic activity and selectivity to the h-GTBE. In this line, Carvalho et al. [70] prepared sulfonated carbons from rice husk by different acid treatments (i.e., using sulfuric acid 6M, concentrated sulfuric acid, and also the vapor of concentrated sulfuric acid). From all the catalysts studied, the sulfonated ones using concentrated sulfuric acid under reflux exhibited the best catalytic behavior (53% of glycerol conversion and 25% of selectivity to h-GTBE), due to the highest acidity that this solid exhibited in comparison to the other sulfonated carbons. Gonçalves et al. [71] prepared sulfonated black carbons from coffee grounds, following a similar procedure to that employed by Carvalho [70]. Thus, by the sulfonation of the carbons with fuming sulfuric acid, an incorporation of 8% of sulfur and an acidity value of 4.2 mmol $H^+$/g, were obtained. The yields to MTBG and h-GTBE obtained with these solids were about 40% and 20%, respectively. Furthermore, the active groups of the acid black carbons exhibited a high stability, allowing them to be used in consecutive reactions. Gonçalves et al. [72] also studied the preparation of carbons by hydrothermal treatment of glycerol, that was obtained in the biodiesel production, and sulfuric acid. The best catalytic activity was obtained on the carbon prepared with a glycerol:sulfuric acid ratio of 1:3, attaining yields to the high ethers around 20%, similar to those obtained over the commercial Amberlyst-15. Furthermore, the catalysts were able to be reused in subsequent reactions (up to eight) without an appreciable loss of activity.

Pico et al. [73,74] studied the capability of several ion exchange resins, A-15, A-200, and Amberlite IRC-50, to be reused, concluding that no significant losses of activity were observed for the spent catalysts. The highest activity was obtained over A-15 ($X_G$ = 80%), and was explained on the basis of its better textural properties in addition to its well-known acidity. A negligible loss of activity was also observed by Magar et al. [75] over several acidic clay catalysts after several uses. Concretely, with a fresh montmorillonite KSF/O clay catalyst, total glycerol conversion was attained due to its highest acidity. Celdeira et al. [76] reported the sulfonation of aluminum pillary clay and niobia (HY-340 CBMM) by two different methods, a treatment with 30% aqueous sulfuric acid solution or a treatment with fuming sulfuric acid, achieving a higher sulfur incorporation with fuming sulfuric acid, attaining a yield to h-GTBE of 38.6 with 95% glycerol conversion. Srinivas et al. [77] prepared cesium exchanged tungstophosphoric acid (CsTPA), supported on tin oxide with cesium exchanges between 10% and 30%. The acidity of the solids increased up to 20% of cesium exchange, being the solid with the highest acidity and also exhibiting the best catalytic performance ($X_G$ = 91% and $S_{h\text{-}GTBE}$ = 44%) after only one hour of reaction time. The authors proposed a plausible mechanism for the reaction, in which Bronsted acid sites on CsTPA were responsible for activating *tert*-Butanol, while Lewis acid sites in the support activated the glycerol, resulting in the formation of mono-ethers. The successive transformation of glycerol to h-GTBE is possible due to the regeneration of acid sites. Furthermore, the authors revealed that these catalysts were also water tolerant.

In addition to the acidity of the catalysts, the textural properties also exhibit a great importance in this reaction, due to the large size of the h-GTBE molecules. Klepàcová et al. [64] compared Amberlyst ion-exchange resins either in gel or macroreticular form, concluding that the macroreticular ones with a high degree of crosslinking exhibited a higher activity due to their larger pores allowing the formation of the TTBG.

The effect of the textural properties on h-GBTE production has also been studied over zeolitic materials. Thus, González et al. [63] studied the effect of hierarchical porosity in a β-zeolite, obtaining a better catalytic behavior than on the conventional zeolite, due to the greatest accessibility of the reactants to the active sites. Furthermore, the hierarchical zeolite after a fluorinated treatment improved the selectivity to h-GTBE, due to the increase in the number of strong acid sites that it exhibited. The decisive role of the Bronsted acidity and porosity was corroborated by the authors [78], studying several type of zeolites, such as mordenite, beta, and ZSM-5 zeolites modified by protonation, dealumination, desilication-protonation, lanthanum exchange, and fluorination. Thus, the best catalytic behavior ($X_G$ = 75% and $S_{h\text{-}GTBE}$ = 37%) was associated with the highest acidity and the larger pore size of the zeolite. Likewise, Simone et al. [79] achieved better results on nanostructured MFI-type zeolites than on traditional zeolites. The nanosponge-like morphology of the zeolites consisted of a three-dimensional disordered network of MFI layers with 2.5 nm thickness supporting each other, obtaining the highest selectivity to TTBG (10%) with unilamellar MFI containing a SI:Al ratio of 100, which exhibited a high proportion of the acid sites on the external surface, enabling a favorable accessibility of the reactants to these sites. Estevez et al. [80] dealuminated a HZSM-5, with different Si:Al ratio, and a HY zeolite by acid treatment with HCl and functionalized them with two organosilica precursors: 3-mercaptopropyltrimethoxysilane (M), containing thiol groups, and 2-(4-chlorosulfonylphenyl)ethyltrimethoxysilane (C), which contains sulfonic acid groups. The solids with highest acidity were those functionalized with M, also exhibiting the highest yields to h-GTBE (13%), using microwave as a way of heating. Furthermore, the textural properties of the zeolites played an important role in their activity (i.e., HY, with the largest channel size, was more active than the HZSM-5). Very recently, Miranda et al. [81] explored the impact of the size of the crystals and the structures of four types of zeolites (i.e., FAU, MOR, BEA, and MFI). They proposed that the reaction followed an Eley–Rideal mechanism in which the total amount of Bronsted acid sites is decisive, but also the shape selectivity. In fact, the formation of DTBGs was rather low in comparison to that obtained on meso- and macroporous acid catalysts. Veiga et al. [82] evaluated the catalytic performance of zeolites subjected to a steaming treatment and acid leaching to eliminate extra-framework aluminum, reporting that the hydrophobicity–hydrophilicity relationship of the catalysts had great influence on the catalytic activity, in addition to their acid and textural properties. The influence of the hydrophobic–hydrophilic character of the catalysts was also claimed by Estevez et al. [83], by studying sulfonated silica-based and sulfonated organosilica-based materials exhibiting different acidities, textural properties, and also different hydrophilic–hydrophobic character. The acidity and hydrophilic character were key parameters for the catalytic behavior. In fact, the best results ($Y_{h\text{-}GTBE}$ = 28%) were obtained over the hybrid silica exhibiting the highest acidity, the $S_{50}TS_{50}O$, at 75 °C and autogenous pressure. In the same line, Estevez et al. [84] reported a yield value to h-GTBE of 21% after 15 min of reaction using microwave as a way of heating in the reaction, on organosilica-aluminum phosphates that exhibited the highest acidity and a similar proportion of meso- and macropores. Furthermore, the hydrophilic character of the solids avoided the rapid deactivation observed on Amberlyst-15.

**Table 2.** Compilation of studies on glycerol etherification with TBA over heterogeneous catalysis.

| Catalyst [a] | Reaction Conditions | | | | | $X_G$ (mol %) | $S_{h\text{-}GTBE}$ (mol %) | Ref. |
|---|---|---|---|---|---|---|---|---|
| | T (°C) | P [b] (bar) | Cat. Loading (wt.% of G) | TBA/G Mol/mol | Time (h) | | | |
| A-15-dry | 60 | Auto | 7.5 | 4 | 8 | 79 | 19 | [39] |
| Hierarchical-Beta zeolite | 75 | Auto | 5 | 4 | 24 | 77 | 35 | [63] |
| A-35-dry | 90 | Auto | 7.5 | 4 | 8 | 69 | 24 | [64] |
| A-70 | 190 | Auto | 7.5–10 | 2 | 8 | 60 | 37 * | [65] |
| A-15 | 70 | 1 | 7.5 | 4 | 6 | 94 | 30 | [66] |
| H730/ES70Y | 80 | Auto | 7.5 | 8 | 27 | 100 | 70 | [67] |
| SCC-S | 120 | Auto | 5 | 4 | 7 | 81 | 21 | [69] |
| TC-L | 120 | Auto | 5 | 4 | 8 | 53 | 25 | [70] |
| BCC-S5h | 120 | Auto | 5 | 4 | 5 | 70 | 29 | [71] |
| BC 1:3 | 120 | Auto | 5 | 4 | 6 | 75 | 29 | [72] |
| A-15 | 60 | Auto | 7.5 | 4 | 8 | 80 | 20 | [73,74] |
| Mont-KSF/O | 110 | Auto | 27 | 20 | 6 | ~100 | ~30 | [75] |
| AS-100 | 120 | Auto | 5 | 4 | 5 | 100 | 40 | [76] |

**Table 2.** *Cont.*

| Catalyst [a] | Reaction Conditions | | | | | $X_G$ (mol %) | $S_{h\text{-}GTBE}$ (mol %) | Ref. |
|---|---|---|---|---|---|---|---|---|
| | T (°C) | P [b] (bar) | Cat. Loading (wt.% of G) | TBA/G Mol/mol | Time (h) | | | |
| $20C_1TS$ | 100 | Auto | 27 | 6 | 1 | 91 | 44 | [77] |
| FHB | 75 | Auto | 5 | 4 | 24 | 75 | 37 | [78] |
| MFI-UL-100 | 120 | Auto | 5 | 4 | 12 | 82 | 24 | [79] |
| M-HY | 85 [c] | Auto | 5 | 4 | 15 min | 59 | 22 | [80] |
| $BEA_{NC}{}^{15}$ | 90 | Auto | 7.5 | 4 | 10 | 57 | 29 | [81] |
| USY-650-L-2 | 90 | Auto | 7.6 | 4 | 4 | 75 | 21 | [82] |
| C(10)AlPO(1.5)-250 | 85 | Auto | 5 | 4 | 15 min | 83 | 25 | [84] |

a. SCC-S: Sulfonated carbon from sugar cane bagasse; TC-L: Sulfonated carbon from rice husk, using concentrated sulfuric acid as the sulfonating agent; BCC-S5h: Sulfonated carbon from coffee-ground waste, using fuming sulfuric acid as the sulfonating agent; BC 1:3: Sulfonated black carbon obtained by hydrothermal treatment of glycerol and sulfuric acid; Mont-KSF/O: Montmorillonite clay catalyst; AS-100: Pillary clay sulfonated with fuming sulfuric acid; $20C_1TS$: Cesium exchanged phosphotungstic acid; H730/ES70Y: Hyflon Ion S4X supported on microspherical silica; FHB: Fluorinated h-Beta zeolite; MFI-UL-100: Unilamellar MFI zeolite with Si:Al ratio of 100; M-HY: HY zeolite modified with 3-mercaptopropyltrimethoxysilane; $BEA_{NC}{}^{15}$: Nanometer size crystal BEA zeolite with Si:Al ratio of 15; USY-650-L-2: USY zeolite after two cycles of heating treatment (650 °C of steaming temperature) and the leaching of acid sites; C(10)AlPO(1.5)-250: Organosilica-aluminum phosphate with a molar ratio Al:P of 1.5, calcined at 250 °C and 10 mmol of 2-(4-chlorosulfonylphenyl)ethyltrimethoxysilane. b Auto: Autogenous pressure. c. Reaction carried out under microwave irradiation. * Selectivity to HCs, HCs being the products from the reaction between G and either IB or its oligomers.

Experimental Designs Reported to Date for Glycerol Etherification with TBA

The results overviewed in Section 2.2 are referred to as the etherification of glycerol with TBA carried out in batch reactors, under different reaction conditions. However, some studies have also been accomplished using continuous flow reactors [25]. Thus, Ozbay et al. [85] compared the results obtained over a batch reactor employing the Amberlyst-15 as a catalyst, with those obtained using a flow reactor, obtaining similar results with both reactors, although much longer reaction times are needed with batch reactors. In addition, the effect of feed composition, reaction temperature, and pressure on the product distribution was also studied. Independently of the type of reactor employed, the generation of water during the reaction was still a problem. Authors adopted the strategy previously employed on a batch reactor, which was the use of Zeolite 4A to remove water from the reaction medium, increasing the di-ethers production, as reported on batch reactors [66]. A catalytic screening in the flow reactor system was reported by Ozbay et al. [86], indicating that despite the fact that the Bronsted acidity was the most important property of the catalysts, the textural properties were also of great importance in order to avoid the diffusion resistance. The importance of textural properties was corroborated by Viswanadham et al. [87] obtaining the best results (95% of glycerol conversion and selectivity to h-GTBE of 99%) on a Nano-Bea zeolite in a continuous flow reactor, due to the presence of inter-crystalline mesopores, which were absent in the typical BEA zeolite. Vlad et al. [88] reported a plant-wide control for this reaction, employing a flow reactor and the Amberlyst-15 as a catalyst. The separation of the h-GTBE from the reaction mixture was achieved by two distillation columns. Authors have concluded that this process is economically feasible at the plant scale only if the mono-ethers can be recirculated for subsequent reaction with TBA. However, the separations of these MTBGs and reactants were difficult because of the formation of a water–TBA azeotrope, although this azeotrope could be broken using a suitable solvent. Thus, the authors evaluated different solvents for the separation of MTBGs, attaining good results with an extractive distillation, using 1,4-butanediol as the solvent [89]. However, due to the large number of units involved, the predicted annual cost was very high. Another option consisting in a reactive distillation column, where *tert*-Butyl alcohol was fed as vapor lower down the reactive zone, was also evaluated, reducing the costs almost 20 times. Very recently, Simasatitkul et al. [90] also studied the techno-economic assessment of extractive distillation for *tert*-Butyl alcohol recovery in the etherification of glycerol, reporting the use of different solvents for breaking the water–TBA azeotrope, concluding that hexyl acetate was the most suitable solvent, based not only on the environmental impact but also on the annual costs and effectiveness.

Kiatkittipong et al. [62] reported simulation and experimental studies about the use of reactive distillation to perform the reaction, finding that the suitable configuration consisted of six rectifying stages and six reaction stages. Likewise, they made a theoretical analysis about the Gibbs free energy and also performed a kinetic study of the reaction process.

Singh et al. [91] reported a system consisting of two interconnected autoclaves. In the first one, the dehydration of TBA was performed and subsequently conducted to the second autoclave, where the etherification of glycerol with IB took place.

Last but not least, the use of microwave irradiation as an alternative to conventional heating for this reaction has been considered by some authors [84,92]. In general, under microwave irradiation, similar results were obtained, although the reaction time to achieve these results was considerably reduced, making this option very promising for the near future.

## 3. Blends of Glycerol Ethers Additives and (Bio)Fuel

Diesel engines have several advantages from the energetic point of view in comparison to explosion engines. However, the combustion of diesel fuel promotes certain environmental problems that must be mitigated, such as the particulate matter emission. The oxygenated fuel additives can diminish the particulate matter and $NO_x$ emissions, since the higher amount of oxygen favors the combustion process. Nevertheless, this amount of oxygen must be controlled, because a high temperature during the combustion can promote a higher oxidation of the $N_2$ of air, thus increasing the $NO_x$ emissions. For considering an oxygenated molecule as a potential additive for fuel formulations, it has to comply with the current regulations (e.g., European Standards (ENs) in the European Union and the American Society for Testing and Materials (ASTM) for United States and Canada) [26]. These standards collect the requirements and test methods for fuel for being employed in engines. Concretely, the EN 14214 and the ASTM D6751 compile the requirements for biodiesel.

To the best of our knowledge, the influence of glycerol ethers as oxygenated additives for (bio)fuel is still poorly studied. Noureddini et al. [93] determined the physical properties of glycerol ether-biodiesel blends, indicating that these additives are soluble in diesel and biodiesel up to 22%. Furthermore, these additives showed a great potential to improve the properties of biodiesel at low temperatures (i.e., the pour point (temperature below which the liquid loss its flow characteristic) and cloud point (temperature below which the liquid forms a cloudy appearance) from 0 °C to –5 °C and from –3 °C to –6 °C, respectively), and exhibiting negligible impact on fuel specific gravity. These additives also decreased the viscosity of biodiesel from 5.9 to 5.4 $mm^2/s$ for a blend of 12% of glycerol ethers and 88% of biodiesel [94]. Melero et al. [95] evaluated different oxygenated compounds (solketal, triacetin, mix of ethers, and mix of esters) in blends with biodiesel, according to the procedures listed in the EN 14214, obtaining the best results for the mixture of ethers. Thus, they improved the properties of biodiesel at low temperature and the viscosity values without affecting other important biodiesel quality parameters.

A deeper study about the behavior of several additives from glycerol (acetals, ethers, carbonates, etc.) corroborated that h-GTBE is the best additive to diminish the particulate matter and $NO_x$ emissions [36]. Furthermore, the physicochemical properties of a blend composed of 5% of biofuel (92.5% of a rapeseed methyl ester + 7.5 of h-GTBE) and 95% of diesel comply with the EN 14214 standard and did not present any technical disadvantage for its use in the engines. In fact, a h-GTBE-diesel blend, in a proportion of 1:10, exhibited lower emissions of soot and $NO_x$ than fossil diesel without affecting the combustion process of different types of engines [51,52,96].

Likewise, the influence of additives such as solketal, h-GTBE, and solketal-*tert*-Butyl ether (obtained by etherification between solketal and TBA) in the anti-wear properties (ASTM D 2266-01) of a low viscosity hydrocarbon oil fraction has been also evaluated [97]. From these additives, the solketal exhibited the best anti-wear properties, improving them by 42%, in respect to the free-additive hydrocarbon oil.

Therefore, in general, the use of oxygenated fuel additives derived from glycerol and, concretely, the glycerol ethers is a matter of the economics of the production process.

## 4. Conclusions

In recent decades, the increase in the production of biodiesel has generated a huge amount of crude glycerol, whose valorization would have a great influence on the biodiesel process, making it more affordable from an economical point of view. In this context, the research of cheaper and more sustainable routes to transform glycerol into value-added products in the presence of heterogeneous catalysts are being widely developed.

Among all the routes being studied, the production of oxygenated additives for fuels has been raised as one of the most promising options. Specifically, the di- and tri-*tert*-Butyl ethers (h-GTBE) obtained by the reaction of glycerol, either with isobutene or *tert*-Butyl alcohol, improve the efficiency of the diesel combustion and reduce the particulate matter and the soot emissions with outstanding results. Furthermore, some studies foresee that these additives could be produced on industrial scale, so the search for suitable heterogeneous acid catalysts for the etherification reaction has been of great interest in recent years.

In general, a good catalyst for this reaction, implying a high yield to h-GTBE, should exhibit not only a high amount of strong acid sites but also textural properties that allow the formation of the products of the reaction, which are molecules with a large volume. In addition, an adequate hydrophobic–hydrophilic character and water-resistance are characteristics also required to avoid the deactivation of the acid sites, mainly when TBA is employed as a reactant. The study of different catalysts has been combined with the study of several engineering designs, in order to improve the production of the desired products. Thus, different alternatives, such as the use of water permselective membranes or any other sorbents of water, as well as different flow reactors have been studied.

Despite the fact that there are catalysts that have shown remarkable results in the etherification with *tert*-Butyl alcohol, the selectivity to h-GTBE can be improved in order to attain similar results than with isobutene. For this purpose, the shift of the reaction equilibrium to the formation of poly-substituted ethers is still a challenge. Thus, researches contemplating catalysts with ideal properties to perform the reaction, as well as new forms to remove the water formed in the reaction, could be a possible way to go forward.

Nevertheless, further research efforts are necessary in order to overcome some of the remaining challenges regarding these reactions. For example, the use of crude glycerol directly obtained from biodiesel production, as the majority of these studies employed pure glycerol as a reactant. Furthermore, the oxygenated fuel additives must be tested in different fuel blends to ensure their behavior, according to specification standards, etc. Likewise, the production of these additives on a larger scale in a feasible way needs to be achieved.

**Author Contributions:** R.E. conceived, reviewed, and wrote the manuscript. L.A.-D. helped in writing the manuscript. D.L. and F.M.B. supervised the manuscript.

**Funding:** This research received no external funding.

**Acknowledgments:** This research is supported by the MEIC funds (Project ENE 2016-81013-R), that cover the costs to publish in open access.

**Conflicts of Interest:** The authors declare no conflicts of interest.

## Abbreviations

| | |
|---|---|
| A-15 | Amberlyst-15 |
| DIB | Di-isobutene |
| DOE | United States Department of Energy |
| DTBGs | Di-*tert*-Butyl glycerol ethers |
| EGR | Exhaust gas recirculation |
| EIA | U.S. Energy Information Administration |
| ETBE | Ethyl *tert*-Butyl ether |
| EN 14214 | Standard published by the European Committee for Standardization that describes the requirements and test methods for FAME |
| FAEE | Fatty acid ethyl ester |
| FAME | Fatty acid methyl esters, components of conventional biodiesel |
| G | Glycerol |
| h-GTBE | "High" glycerol *tert*-Butyl ethers (i.e., DTBGs + TTBG) |
| IB | Isobutene |
| MTBE | Methyl *tert*-Butyl ether |
| MTBGs | Methyl *tert*-Butyl glycerol ethers |
| OPEC | Organization of the Petroleum Exporting Countries |
| TBA | *tert*-Butyl alcohol |
| TTBG | Tri-*tert*-Butyl glycerol ether |
| VLE | Vapor liquid equilibrium |

## References

1. Available online: https://countryeconomy.com/raw-materials/opec (accessed on 17 June 2019).
2. Nda-Umar, U.; Ramli, I.; Taufiq-Yap, Y.; Muhamad, E. An Overview of Recent Research in the Conversion of Glycerol into Biofuels, Fuel Additives and other Bio-Based Chemicals. *Catalysts* **2019**, *9*, 15. [CrossRef]
3. Hurtado, B.; Posadillo, A.; Luna, D.; Bautista, F.; Hidalgo, J.; Luna, C.; Calero, J.; Romero, A.; Estevez, R. Synthesis, Performance and Emission Quality Assessment of Ecodiesel from Castor Oil in Diesel/Biofuel/Alcohol Triple Blends in a Diesel Engine. *Catalysts* **2019**, *9*, 40. [CrossRef]
4. Chum, H.L.; Overend, R.P. Biomass and bioenergy in the United States. *Adv. Sol. Energy* **2003**, *15*, 83–148.
5. Akia, M.; Yazdani, F.; Motaee, E.; Han, D.; Arandiyan, H. A review on conversion of biomass to biofuel by nanocatalysts. *Biofuel Res. J.* **2014**, *1*, 16–25. [CrossRef]
6. Huber, G.W.; Corma, A. Synergien zwischen Bio-und Ölraffinerien bei der Herstellung von Biomassetreibstoffen. *Angew. Chem.* **2007**, *119*, 7320–7338. [CrossRef]
7. Corma, A.; Iborra, S.; Velty, A. Chemical routes for the transformation of biomass into chemicals. *Chem. Rev.* **2007**, *107*, 2411–2502. [CrossRef] [PubMed]
8. Pagliaro, M.; Ciriminna, R.; Kimura, H.; Rossi, M.; Della Pina, C. From Glycerol to Value-Added Products. *Angew. Chem. Int. Ed.* **2007**, *46*, 4434–4440. [CrossRef]
9. Sheldon, R.A. Green and sustainable manufacture of chemicals from biomass: State of the art. *Green Chem.* **2014**, *16*, 950–963. [CrossRef]
10. Cornejo, A.; Barrio, I.; Campoy, M.; Lázaro, J.; Navarrete, B. Oxygenated fuel additives from glycerol valorization. Main production pathways and effects on fuel properties and engine performance: A critical review. *Renew. Sustain. Energy Rev.* **2017**, *79*, 1400–1413. [CrossRef]
11. Ragauskas, A.J.; Williams, C.K.; Davison, B.H.; Britovsek, G.; Cairney, J.; Eckert, C.A.; Frederick, W.J.; Hallett, J.P.; Leak, D.J.; Liotta, C.L. The path forward for biofuels and biomaterials. *Science* **2006**, *311*, 484–489. [CrossRef]
12. Mohan, S.V.; Nikhil, G.; Chiranjeevi, P.; Reddy, C.N.; Rohit, M.; Kumar, A.N.; Sarkar, O. Waste biorefinery models towards sustainable circular bioeconomy: Critical review and future perspectives. *Bioresour. Technol.* **2016**, *215*, 2–12. [CrossRef]
13. Monthly Biodiesel Production Report. Available online: https://www.eia.gov/biofuels/biodiesel/production/ (accessed on 17 June 2019).
14. REN21. Renewable Energy Policy Network for the 21st Century. Renewables 2018 Global Status Report. Available online: http://www.ren21.net/wp-content/uploads/2018/06/17-8652_GSR2018_FullReport_web_final_.pdf (accessed on 17 June 2019).

15. Ridjan, I.; Mathiesen, B.V.; Connolly, D.; Duić, N. The feasibility of synthetic fuels in renewable energy systems. *Energy* **2013**, *57*, 76–84. [CrossRef]
16. Kumar, A.; Sharma, S. Potential non-edible oil resources as biodiesel feedstock: An Indian perspective. *Renew. Sustain. Energy Rev.* **2011**, *15*, 1791–1800. [CrossRef]
17. Knothe, G.; Razon, L.F. Biodiesel fuels. *Prog. Energy Combust. Sci.* **2017**, *58*, 36–59. [CrossRef]
18. Borges, M.E.; Díaz, L. Recent developments on heterogeneous catalysts for biodiesel production by oil esterification and transesterification reactions: A review. *Renew. Sustain. Energy Rev.* **2012**, *16*, 2839–2849. [CrossRef]
19. Avhad, M.; Marchetti, J. A review on recent advancement in catalytic materials for biodiesel production. *Renew. Sustain. Energy Rev.* **2015**, *50*, 696–718. [CrossRef]
20. Ramachandran, K.; Suganya, T.; Gandhi, N.N.; Renganathan, S. Recent developments for biodiesel production by ultrasonic assist transesterification using different heterogeneous catalyst: A review. *Renew. Sustain. Energy Rev.* **2013**, *22*, 410–418. [CrossRef]
21. Monteiro, M.R.; Kugelmeier, C.L.; Pinheiro, R.S.; Batalha, M.O.; da Silva César, A. Glycerol from biodiesel production: Technological paths for sustainability. *Renew. Sustain. Energy Rev.* **2018**, *88*, 109–122. [CrossRef]
22. Ayoub, M.; Abdullah, A.Z. Critical review on the current scenario and significance of crude glycerol resulting from biodiesel industry towards more sustainable renewable energy industry. *Renew. Sustain. Energy Rev.* **2012**, *16*, 2671–2686. [CrossRef]
23. Comelli, R.A. Glycerol, the co-product of biodiesel: One key for the future bio-refinery. In *Biodiesel-Quality, Emissions and By-Products*; IntechOpen: London, UK, 2011.
24. Luo, X.; Ge, X.; Cui, S.; Li, Y. Value-added processing of crude glycerol into chemicals and polymers. *Bioresour. Technol.* **2016**, *215*, 144–154. [CrossRef]
25. Len, C.; Delbecq, F.; Corpas, C.C.; Ramos, E.R. Continuous flow conversion of glycerol into chemicals: An overview. *Synthesis* **2018**, *50*, 723–741. [CrossRef]
26. Rahmat, N.; Abdullah, A.Z.; Mohamed, A.R. Recent progress on innovative and potential technologies for glycerol transformation into fuel additives: A critical review. *Renew. Sustain. Energy Rev.* **2010**, *14*, 987–1000. [CrossRef]
27. Bagheri, S.; Julkapli, N.M.; Yehye, W.A. Catalytic conversion of biodiesel derived raw glycerol to value added products. *Renew. Sustain. Energy Rev.* **2015**, *41*, 113–127. [CrossRef]
28. Nanda, M.R.; Zhang, Y.; Yuan, Z.; Qin, W.; Ghaziaskar, H.S.; Xu, C.C. Catalytic conversion of glycerol for sustainable production of solketal as a fuel additive: A review. *Renew. Sustain. Energy Rev.* **2016**, *56*, 1022–1031. [CrossRef]
29. Bozkurt, Ö.D.; Tunc, F.M.; Bağlar, N.; Celebi, S.; Günbaş, İ.D.; Uzun, A. Alternative fuel additives from glycerol by etherification with isobutene: Structure–performance relationships in solid catalysts. *Fuel Process. Technol.* **2015**, *138*, 780–804. [CrossRef]
30. Izquierdo, J.; Montiel, M.; Palés, I.; Outón, P.; Galán, M.; Jutglar, L.; Villarrubia, M.; Izquierdo, M.; Hermo, M.; Ariza, X. Fuel additives from glycerol etherification with light olefins: State of the art. *Renew. Sustain. Energy Rev.* **2012**, *16*, 6717–6724. [CrossRef]
31. Yee, K.F.; Mohamed, A.R.; Tan, S.H. A review on the evolution of ethyl tert-Butyl ether (ETBE) and its future prospects. *Renew. Sustain. Energy Rev.* **2013**, *22*, 604–620. [CrossRef]
32. Garcia, E.; Laca, M.; Pérez, E.; Garrido, A.; Peinado, J. New class of acetal derived from glycerin as a biodiesel fuel component. *Energy Fuels* **2008**, *22*, 4274–4280. [CrossRef]
33. Oprescu, E.-E.; Stepan, E.; Dragomir, R.E.; Radu, A.; Rosca, P. Synthesis and testing of glycerol ketals as components for diesel fuel. *Fuel Process. Technol.* **2013**, *110*, 214–217. [CrossRef]
34. Zare, A.; Nabi, M.N.; Bodisco, T.A.; Hossain, F.M.; Rahman, M.M.; Ristovski, Z.D.; Brown, R.J. The effect of triacetin as a fuel additive to waste cooking biodiesel on engine performance and exhaust emissions. *Fuel* **2016**, *182*, 640–649. [CrossRef]
35. Alptekin, E. Emission, injection and combustion characteristics of biodiesel and oxygenated fuel blends in a common rail diesel engine. *Energy* **2017**, *119*, 44–52. [CrossRef]
36. Jaecker-Voirol, A.; Durand, I.; Hillion, G.; Delfort, B.; Montagne, X. Glycerin for new biodiesel formulation. *Oil Gas Sci. Technol. Rev. l'IFP* **2008**, *63*, 395–404. [CrossRef]
37. Behr, A.; Obendorf, L. Development of a Process for the Acid-Catalyzed Etherification of Glycerine and Isobutene Forming Glycerine Tertiary Butyl Ethers. *Eng. Life Sci.* **2002**, *2*, 185–189. [CrossRef]

38. Karinen, R.; Krause, A. New biocomponents from glycerol. *Appl. Catal. A Gen.* **2006**, *306*, 128–133. [CrossRef]
39. Klepáčová, K.; Mravec, D.; Bajus, M. tert-Butylation of glycerol catalysed by ion-exchange resins. *Appl. Catal. A Gen.* **2005**, *294*, 141–147. [CrossRef]
40. Klepáčová, K.; Mravec, D.; Kaszonyi, A.; Bajus, M. Etherification of glycerol and ethylene glycol by isoButylene. *Appl. Catal. A Gen.* **2007**, *328*, 1–13. [CrossRef]
41. Lee, H.J.; Seung, D.; Jung, K.S.; Kim, H.; Filimonov, I.N. Etherification of glycerol by isoButylene: Tuning the product composition. *Appl. Catal. A Gen.* **2010**, *390*, 235–244. [CrossRef]
42. Lee, H.J.; Seung, D.; Filimonov, I.N.; Kim, H. Etherification of glycerol by isoButylene. Effects of the density of acidic sites in ion-exchange resin on the distribution of products. *Korean J. Chem. Eng.* **2011**, *28*, 756–762. [CrossRef]
43. Bozkurt, Ö.D.; Bağlar, N.; Çelebi, S.; Uzun, A. Assessment of acid strength in sodium-exchanged resin catalysts: Consequences on glycerol etherification with isobutene in batch and flow reactors. *Mol. Catal.* **2019**, *466*, 1–12. [CrossRef]
44. Melero, J.; Vicente, G.; Morales, G.; Paniagua, M.; Moreno, J.; Roldán, R.; Ezquerro, A.; Pérez, C. Acid-catalyzed etherification of bio-glycerol and isoButylene over sulfonic mesostructured silicas. *Appl. Catal. A Gen.* **2008**, *346*, 44–51. [CrossRef]
45. Zhao, W.; Yi, C.; Yang, B.; Hu, J.; Huang, X. Etherification of glycerol and isoButylene catalyzed over rare earth modified Hβ-zeolite. *Fuel Process. Technol.* **2013**, *112*, 70–75. [CrossRef]
46. Xiao, L.; Mao, J.; Zhou, J.; Guo, X.; Zhang, S. Enhanced performance of HY zeolites by acid wash for glycerol etherification with isobutene. *Appl. Catal. A Gen.* **2011**, *393*, 88–95. [CrossRef]
47. González, M.D.; Salagre, P.; Taboada, E.; Llorca, J.; Cesteros, Y. Microwave-assisted synthesis of sulfonic acid-functionalized microporous materials for the catalytic etherification of glycerol with isobutene. *Green Chem.* **2013**, *15*, 2230–2239. [CrossRef]
48. González, M.D.; Cesteros, Y.; Llorca, J.; Salagre, P. Boosted selectivity toward high glycerol tertiary Butyl ethers by microwave-assisted sulfonic acid-functionalization of SBA-15 and beta zeolite. *J. Catal.* **2012**, *290*, 202–209. [CrossRef]
49. González, M.D.; Salagre, P.; Mokaya, R.; Cesteros, Y. Tuning the acidic and textural properties of ordered mesoporous silicas for their application as catalysts in the etherification of glycerol with isobutene. *Catal. Today* **2014**, *227*, 171–178. [CrossRef]
50. González, M.D.; Salagre, P.; Taboada, E.; Llorca, J.; Molins, E.; Cesteros, Y. Sulfonic acid-functionalized aerogels as high resistant to deactivation catalysts for the etherification of glycerol with isobutene. *Appl. Catal. B Environ.* **2013**, *136*, 287–293. [CrossRef]
51. Frusteri, F.; Cannilla, C.; Bonura, G.; Spadaro, L.; Mezzapica, A.; Beatrice, C.; Di Blasio, G.; Guido, C. Glycerol ethers production and engine performance with diesel/ethers blend. *Top. Catal.* **2013**, *56*, 378–383. [CrossRef]
52. Beatrice, C.; Di Blasio, G.; Lazzaro, M.; Cannilla, C.; Bonura, G.; Frusteri, F.; Asdrubali, F.; Baldinelli, G.; Presciutti, A.; Fantozzi, F. Technologies for energetic exploitation of biodiesel chain derived glycerol: Oxy-fuels production by catalytic conversion. *Appl. Energy* **2013**, *102*, 63–71. [CrossRef]
53. Frusteri, F.; Frusteri, L.; Cannilla, C.; Bonura, G. Catalytic etherification of glycerol to produce biofuels over novel spherical silica supported Hyflon®catalysts. *Bioresour. Technol.* **2012**, *118*, 350–358. [CrossRef] [PubMed]
54. Zhao, W.; Yang, B.; Yi, C.; Lei, Z.; Xu, J. Etherification of glycerol with isoButylene to produce oxygenate additive using sulfonated peanut shell catalyst. *Ind. Eng. Chem. Res.* **2010**, *49*, 12399–12404. [CrossRef]
55. Zhou, J.; Wang, Y.; Guo, X.; Mao, J.; Zhang, S. Etherification of glycerol with isobutene on sulfonated graphene: Reaction and separation. *Green Chem.* **2014**, *16*, 4669–4679. [CrossRef]
56. Voicu, V.; Bombos, D.; Bolocan, I.; Jang, C.R.; Ciuparu, D. The Influence of the Character of Emulsifiers on the Performance of H (4) SiW (12) O (40 center dot) 30H (2) O Heteropolyacid Catalyst in Glycerol Etherification with Isobutene. *Rev. Chim.* **2012**, *63*, 200–204.
57. Liu, J.; Yuan, Y.; Pan, Y.; Huang, Z.; Yang, B. Liquid–liquid equilibrium for systems of glycerol and glycerol tert-Butyl ethers. *Fluid Phase Equilibria* **2014**, *365*, 50–57. [CrossRef]
58. Liu, J.; Yang, B.; Yi, C. Kinetic study of glycerol etherification with isobutene. *Ind. Eng. Chem. Res.* **2013**, *52*, 3742–3751. [CrossRef]
59. Di Serio, M.; Cozzolino, M.; Giordano, M.; Tesser, R.; Patrono, P.; Santacesaria, E. From homogeneous to heterogeneous catalysts in biodiesel production. *Ind. Eng. Chem. Res.* **2007**, *46*, 6379–6384. [CrossRef]

60. Liu, J.; Daoutidis, P.; Yang, B. Process design and optimization for etherification of glycerol with isobutene. *Chem. Eng. Sci.* **2016**, *144*, 326–335. [CrossRef]
61. Turan, A.; Hrivnák, M.; Klepáčová, K.; Kaszonyi, A.; Mravec, D. Catalytic etherification of bioglycerol with C4 fraction. *Appl. Catal. A Gen.* **2013**, *468*, 313–321. [CrossRef]
62. Kiatkittipong, W.; Intaracharoen, P.; Laosiripojana, N.; Chaisuk, C.; Praserthdam, P.; Assabumrungrat, S. Glycerol ethers synthesis from glycerol etherification with tert-Butyl alcohol in reactive distillation. *Comput. Chem. Eng.* **2011**, *35*, 2034–2043. [CrossRef]
63. González, M.D.; Salagre, P.; Linares, M.; García, R.; Serrano, D.; Cesteros, Y. Effect of hierarchical porosity and fluorination on the catalytic properties of zeolite beta for glycerol etherification. *Appl. Catal. A Gen.* **2014**, *473*, 75–82. [CrossRef]
64. Klepáčová, K.; Mravec, D.; Bajus, M. Etherification of glycerol with tert-Butyl alcohol catalysed by ion-exchange resins. *Chem. Pap.* **2006**, *60*, 224–230. [CrossRef]
65. Chang, J.-S.; Zhang, Y.-C.; Chen, C.-C.; Ling, T.-R.; Chiou, Y.-J.; Wang, G.-B.; Chang, K.-T.; Chou, T.-C. One-step synthesis of gasoline octane booster and diesel fuel from glycerol and tert-Butyl alcohol. *Ind. Eng. Chem. Res.* **2014**, *53*, 5398–5405. [CrossRef]
66. Frusteri, F.; Arena, F.; Bonura, G.; Cannilla, C.; Spadaro, L.; Di Blasi, O. Catalytic etherification of glycerol by tert-Butyl alcohol to produce oxygenated additives for diesel fuel. *Appl. Catal. A Gen.* **2009**, *367*, 77–83. [CrossRef]
67. Cannilla, C.; Bonura, G.; Frusteri, L.; Frusteri, F. Glycerol Etherification with TBA: High Yield to Poly-Ethers Using a Membrane Assisted Batch Reactor. *Environ. Sci. Technol.* **2014**, *48*, 6019–6026. [CrossRef] [PubMed]
68. Cannilla, C.; Bonura, G.; Frusteri, L.; Frusteri, F. Catalytic production of oxygenated additives by glycerol etherification. *Open Chem.* **2014**, *12*, 1248–1254. [CrossRef]
69. Gonçalves, M.; Souza, V.C.; Galhardo, T.S.; Mantovani, M.; Figueiredo, F.v.C.; Mandelli, D.; Carvalho, W.A. Glycerol conversion catalyzed by carbons prepared from agroindustrial wastes. *Ind. Eng. Chem. Res.* **2013**, *52*, 2832–2839. [CrossRef]
70. Galhardo, T.S.; Simone, N.; Gonçalves, M.; Figueiredo, F.C.; Mandelli, D.; Carvalho, W.A. Preparation of sulfonated carbons from rice husk and their application in catalytic conversion of glycerol. *ACS Sustain. Chem. Eng.* **2013**, *1*, 1381–1389. [CrossRef]
71. Gonçalves, M.; Soler, F.C.; Isoda, N.; Carvalho, W.A.; Mandelli, D.; Sepúlveda, J. Glycerol conversion into value-added products in presence of a green recyclable catalyst: Acid black carbon obtained from coffee ground wastes. *J. Taiwan Inst. Chem. Eng.* **2016**, *60*, 294–301. [CrossRef]
72. Gonçalves, M.; Mantovani, M.; Carvalho, W.A.; Rodrigues, R.; Mandelli, D.; Albero, J.S. Biodiesel wastes: An abundant and promising source for the preparation of acidic catalysts for utilization in etherification reaction. *Chem. Eng. J.* **2014**, *256*, 468–474. [CrossRef]
73. Pico, M.P.; Rosas, J.M.; Rodríguez, S.; Santos, A.; Romero, A. Glycerol etherification over acid ion exchange resins: Effect of catalyst concentration and reusability. *J. Chem. Technol. Biotechnol.* **2013**, *88*, 2027–2038. [CrossRef]
74. Pico, M.P.; Romero, A.; Rodríguez, S.; Santos, A. Etherification of glycerol by tert-Butyl alcohol: Kinetic model. *Ind. Eng. Chem. Res.* **2012**, *51*, 9500–9509. [CrossRef]
75. Magar, S.; Kamble, S.; Mohanraj, G.T.; Jana, S.K.; Rode, C. Solid-Acid-Catalyzed Etherification of Glycerol to Potential Fuel Additives. *Energy Fuels* **2017**, *31*, 12272–12277. [CrossRef]
76. Celdeira, P.A.; Goncalves, M.; Figueiredo, F.C.; Dal Bosco, S.M.; Mandelli, D.; Carvalho, W.A. Sulfonated niobia and pillared clay as catalysts in etherification reaction of glycerol. *Appl. Catal. A Gen.* **2014**, *478*, 98–106. [CrossRef]
77. Srinivas, M.; Raveendra, G.; Parameswaram, G.; Prasad, P.S.; Lingaiah, N. Cesium exchanged tungstophosphoric acid supported on tin oxide: An efficient solid acid catalyst for etherification of glycerol with tert-butanol to synthesize biofuel additives. *J. Mol. Catal. A Chem.* **2016**, *413*, 7–14. [CrossRef]
78. González, M.D.; Cesteros, Y.; Salagre, P. Establishing the role of Brønsted acidity and porosity for the catalytic etherification of glycerol with tert-butanol by modifying zeolites. *Appl. Catal. A Gen.* **2013**, *450*, 178–188. [CrossRef]
79. Simone, N.; Carvalho, W.A.; Mandelli, D.; Ryoo, R. Nanostructured MFI-type zeolites as catalysts in glycerol etherification with tert-Butyl alcohol. *J. Mol. Catal. A Chem.* **2016**, *422*, 115–121. [CrossRef]

80. Estevez, R.; Iglesias, I.; Luna, D.; Bautista, F.M. Sulfonic Acid Functionalization of Different Zeolites and Their Use as Catalysts in the Microwave-Assisted Etherification of Glycerol with tert-Butyl Alcohol. *Molecules* **2017**, *22*, 2206. [CrossRef]
81. Miranda, C.; Urresta, J.; Cruchade, H.; Tran, A.; Benghalem, M.; Astafan, A.; Gaudin, P.; Daou, T.; Ramírez, A.; Pouilloux, Y. Exploring the impact of zeolite porous voids in liquid phase reactions: The case of glycerol etherification by tert-Butyl alcohol. *J. Catal.* **2018**, *365*, 249–260. [CrossRef]
82. Veiga, P.M.; Gomes, A.C.; Veloso, C.O.; Henriques, C.A. Acid zeolites for glycerol etherification with ethyl alcohol: Catalytic activity and catalyst properties. *Appl. Catal. A Gen.* **2017**, *548*, 2–15. [CrossRef]
83. Estevez, R.; López, M.; Jiménez-Sanchidrián, C.; Luna, D.; Romero-Salguero, F.; Bautista, F. Etherification of glycerol with tert-Butyl alcohol over sulfonated hybrid silicas. *Appl. Catal. A Gen.* **2016**, *526*, 155–163. [CrossRef]
84. Estevez, R.; Lopez-Pedrajas, S.; Luna, D.; Bautista, F. Microwave-assisted etherification of glycerol with tert-Butyl alcohol over amorphous organosilica-aluminum phosphates. *Appl. Catal. B Environ.* **2017**, *213*, 42–52. [CrossRef]
85. Ozbay, N.; Oktar, N.; Dogu, G.; Dogu, T. Effects of sorption enhancement and isobutene formation on etherification of glycerol with tert-Butyl alcohol in a flow reactor. *Ind. Eng. Chem. Res.* **2011**, *51*, 8788–8795. [CrossRef]
86. Ozbay, N.; Oktar, N.; Dogu, G.; Dogu, T. Activity comparison of different solid acid catalysts in etherification of glycerol with tert-Butyl alcohol in flow and batch reactors. *Top. Catal.* **2013**, *56*, 1790–1803. [CrossRef]
87. Viswanadham, N.; Saxena, S.K. Etherification of glycerol for improved production of oxygenates. *Fuel* **2013**, *103*, 980–986. [CrossRef]
88. Vlad, E.; Bildea, C.S.; Bozga, G. Design and control of glycerol-tert-Butyl alcohol etherification process. *Sci. World J.* **2012**, *2012*, 180617. [CrossRef] [PubMed]
89. Vlad, E.; Bildea, C.S. Reactive Distillation-a Viable Solution for Etherification of Glycerol with tert-Butyl Alcohol. *Chem. Eng.* **2012**, *29*. [CrossRef]
90. Simasatitkul, L.; Kaewwisetkul, P.; Arpornwichanop, A. Techno-economic assessment of extractive distillation for tert-Butyl alcohol recovery in fuel additive production. *Chem. Eng. Process. Process Intensif.* **2017**, *122*, 161–171. [CrossRef]
91. Singh, J.; Kumar, J.; Negi, M.; Bangwal, D.; Kaul, S.; Garg, M. Kinetics and modeling study on etherification of glycerol using isoButylene by in situ production from tert-Butyl alcohol. *Ind. Eng. Chem. Res.* **2015**, *54*, 5213–5219. [CrossRef]
92. Luque, R.; Budarin, V.; Clark, J.H.; Macquarrie, D.J. Glycerol transformations on polysaccharide derived mesoporous materials. *Appl. Catal. B Environ.* **2008**, *82*, 157–162. [CrossRef]
93. Noureddini, H.; Dailey, W.R.; Hunt, B.A. Production of ethers of glycerol from crude glycerol-the by-product of biodlesel production. *Pap. Biomater.* **1998**, 18.
94. Noureddini, H. Process for producing biodiesel fuel with reduced viscosity and a cloud point below thirty-two (32) degrees Fahrenheit. U.S. Patent 6,015,440, 18 January 2000.
95. Melero, J.A.; Vicente, G.; Morales, G.; Paniagua, M.; Bustamante, J. Oxygenated compounds derived from glycerol for biodiesel formulation: Influence on EN 14214 quality parameters. *Fuel* **2010**, *89*, 2011–2018. [CrossRef]
96. Beatrice, C.; Di Blasio, G.; Lazzaro, M.; Mancaruso, E.; Marialto, R.; Sequino, L.; Vaglieco, B.M. Investigation of the combustion in both metal and optical diesel engines using high-glycerol ethers/diesel blends. *Int. J. Engine Res.* **2015**, *16*, 38–51. [CrossRef]
97. Samoilov, V.O.; Ramazanov, D.N.; Nekhaev, A.I.; Maximov, A.L.; Bagdasarov, L.N. Heterogeneous catalytic conversion of glycerol to oxygenated fuel additives. *Fuel* **2016**, *172*, 310–319. [CrossRef]

Article

# Hydrogen Photo-Production from Glycerol Using Nickel-Doped $TiO_2$ Catalysts: Effect of Catalyst Pre-Treatment

**Jesús Hidalgo-Carrillo *, Juan Martín-Gómez, Julia Morales, Juan Carlos Espejo, Francisco José Urbano and Alberto Marinas**

Departamento de Química Orgánica, Instituto Universitario de Investigación en Química Finay Nanoquímica (IUNAN), Universidad de Córdoba, E-14071 Córdoba, Spain

* Correspondence: jesus.hidalgo@uco.es

Received: 20 July 2019; Accepted: 30 August 2019; Published: 30 August 2019

**Abstract:** In the present piece of research, hydrogen production via the photo-reforming of glycerol (a byproduct from biodiesel generation) is studied. Catalysts consisted of titania modified by Ni (0.5% by weight) obtained through deposition–precipitation or impregnation synthetic methods (labelled as Ni-0.5-DP and Ni-0.5-IMP, respectively). Reactions were performed both under UV and solar irradiation. Activity significantly improved in the presence of Ni, especially under solar irradiation. Moreover, pre-reduced solids exhibited higher catalytic activities than untreated solids, despite the "in-situ" reduction of nickel species and the elimination of surface chlorides under reaction conditions (as evidenced by XPS). It is possible that the catalyst pretreatment at 400 °C under hydrogen resulted in some strong metal–support interactions. In summary, the highest hydrogen production value (ca. 2600 micromole $H_2{\cdot}g^{-1}$) was achieved with pre-reduced Ni-0.5-DP solid using UV light for an irradiation time of 6 h. This value represents a 15.7-fold increase as compared to Evonik P25.

**Keywords:** hydrogen production; photo-reforming; glycerol; $Ni/TiO_2$

## 1. Introduction

Fossil fuel depletion and environmental concerns have resulted in the search for clean energies, with one alternative being hydrogen [1]. Its use has two main advantages [2,3]: i) a high chemical energy per mass (120 KJ/g), superior to that of many fossil fuels, and ii) its combustion only results in water; therefore, it does not emit any toxic substance or greenhouse gas into the atmosphere.

Nevertheless, hydrogen does not exist in nature in its molecular $H_2$ form but combined to other elements; thus, it requires dedicated methods for its production. Therefore, whether or not the use of hydrogen as an energy vector can be termed as "fully green" is dependent on its method of production.

Currently, the most widespread hydrogen production methodologies are hydrocarbon reforming with water vapor and water electrolysis. Hydrocarbon reforming has the disadvantages of being based on raw materials which are taken from non-renewable fossil sources and, therefore, the co-generated $CO_2$ directly impacts the environment by the greenhouse effect. An additional drawback of hydrocarbon reforming with water vapor is its high operating temperature. On the other hand, regarding the production of hydrogen through water electrolysis, its main associated problem is the high consumption of electrical energy to carry out the process. Solar thermal energy can be used as an alternative for water electrolysis but, in this case, large and expensive facilities are needed.

Recently, in addition to the mentioned technologies, innovative techniques have been being developed and could be complementary to those already existing in the medium-term future. Some of these hydrogen production techniques are plasma technology [4], biological production methods such as dark fermentation [5,6] or the photocatalytic reforming of oxygenated organic compounds [7,8].

The photocatalytic reforming of oxygenated organic compounds consists in the treatment of these compounds with light radiation in the presence of water, at room temperature and under anaerobic conditions, to generate gaseous hydrogen and carbon dioxide. The potential of hydrogen production through photocatalytic reforming is fulfilled when biomass residues (bio-glycerol or glucose, among others) are used as oxygenated organic compounds since, in this case, the generated $CO_2$ was previously consumed by the biomass during its growth, so there is no net emission of $CO_2$ but a closing of the carbon cycle [9]. In the process, light is used to activate a semiconductor, promoting electrons from the valence to the conduction band. The oxygenated organic compound is used as a sacrificial agent to favor the elimination of the positively-charged holes, whereas electrons are used to reduce protons and generate $H_2$. As for the sacrificial agents, glycerol is an excellent candidate since it is a by-product of biodiesel production [10].

One of the keys to the success of this emergent technology is the development of suitable catalysts (i.e., semiconductors) which are able to maximize light harvesting and therefore the hydrogen production [11]. $TiO_2$ is the most widely used semiconductor as a result of its high photocatalytic activity and due to the fact that it is inexpensive, not toxic and biologically and chemically inert [12]. Its main drawback is its band gap value (ca. 3.2 eV), which means that only ca. 5% of solar irradiation is absorbed. Furthermore, it also exhibits a high electron–hole recombination rate, which is detrimental to the photocatalytic activity [13].

One alternative to overcome these problems is the incorporation of metals to the semiconductor [14] (Figure 1), which can shift the absorption to the visible light and also act as electron traps, thus preventing electron–hole recombination. Noble metals such as silver [15], gold [16], platinum [17] or palladium [18] have been found to be particularly effective, although there is a need to implement some more cost-effective transition metals such as iron [19], nickel [20,21] and copper [22,23].

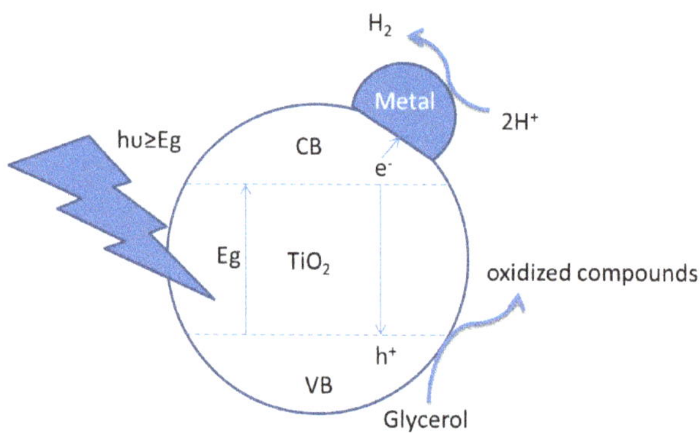


Glycerol photo-reforming: $C_3H_8O_3 + 3H_2O \rightarrow 3CO_2 + 7H_2$

**Figure 1.** Activation of titania using a metal as a co-catalyst.

In the present piece of research, nickel structural (particle size) and chemical properties (oxidation state) in Ni-modified titania photocatalysts has been addressed, and their influence on hydrogen production from glycerol photo-reforming studied. Two catalyst synthetic methods (impregnation vs. deposition–precipitation) and catalyst pre-reduction treatment were analyzed and their influence on the amount of hydrogen photo-produced revealed under both UV and solar radiation.

## 2. Materials and Methods

### 2.1. Catalysts Synthesis

#### 2.1.1. Deposition–precipitation Method

The deposition-precipiation method was used for incorporation of nickel on P25 Evonik with a nominal weight of 0.5%. The incorporation of Nickel was carried out in a "Contalab" system (Switzerland) enabling careful control of all the variables, i.e., pH, temperature, stirrer speed, reactant feed flow, etc. Firstly in 150 mL distilled water were dispersed 5 g of P25 Evonik using ultrasound. The deposition of the nickel was carried out at 60 °C and a pH of 6.8, keeping it constant with an 0.2 M aqueous solution of $K_2CO_3$ (Sigma Aldrich). Next, an aqueous solution of 2 g·$L^{-1}$ of nickel (using $NiCl_2$ (Sigma Aldrich)) was added dropwise and the mixture was maintained for 1 h at 60 °C. Then the solid obtained was filtered and washed with distilled water. Then, the solid was dried at 110 °C overnight. Finally, the solids were calcined at 400 °C for 6 h (using a ramp rate of 10 °C·$min^{-1}$).

#### 2.1.2. Impregnation Method

A total of 5 g of $TiO_2$ (P25 Evonik) was dispersed in 100 mL of distilled water containing the required amount of $NiCl_2$ (Sigma Aldrich) aqueous solution in order to have a nickel nominal content of 0.5% by weight. The suspension was stirred for 1 h and then vacuum filtered at 100 °C. Then, the solid was dried overnight (100 °C) and calcined at 400 °C for 6 h (ramp rate 10 °C·$min^{-1}$).

The catalyst nomenclature includes the metal (Ni), the nominal content (0.5% by weight) and a suffix indicating the synthetic method (DP or IMP for deposition–precipitation or impregnation, respectively).

#### 2.1.3. Catalyst Pre-Reduction

The catalysts were used either as synthesized or after a pre-reduction treatment. In the latter case, the solids were treated at room temperature under $N_2$ flow (20 mL·$min^{-1}$) for 15 min and then submitted to hydrogen (10mL·$min^{-1}$), the temperature being ramped up to 400 °C (rate, 10 °C·$min^{-1}$) and the final temperature being maintained for 1 h. The nomenclature of pre-reduced solids includes the "Red" suffix.

### 2.2. Characterization of the Solids

The determination of the metallic content in the samples was carried out by by inductively coupled plasma mass spectrometry (ICP-MS) using a Perkin Elmer NexionX instrument. Digestion of the samples consisted in dissolution of the sample in an acid solution of 1:1 $H_2O/H_2SO_4$/HF solution at 80 °C and and after in 1:3 $HNO_3$/HCl mixture. The measurements were carried out by the staff at the Central Service for Research Support (SCAI) of the University of Córdoba.

Transmission electron microscopy (TEM) images were obtained with a JEOL JEM 1400 transmission electron microscope. For the measurements were used a 3 mm holey carbon copper grids. Particle sizes were obtained using using the software ImageJ (a public domain Java image processing and analysis program). TEM was carried out at the Central Service for Research Support (SCAI) of the University of Córdoba.

X-ray diffraction (XRD) analysis was performed on a D8 Discover instrument (Bruker Corporation, Billerica, USA) using CuKα radiation over the range 5–80°.

A Cary 1E (Varian) instrument was used for determination of Band Gap values by Diffuse reflectance UV–vis spectra, using as reference material the polytetraethylene (density = 1 g·$cm^{-3}$ and thickness = 6 mm). The plot of the modified Kubelka–Munk function $[F(R)\cdot E]^{1/2}$ versus the energy of the absorbed light E was use to obtain the value of band gap, extrapolating to y = 0 of the linear regression range.

X-ray photoelectron spectroscopy (XPS) data were recorded by a Leibold–Heraeus LHS10 spectrometer capable of operating down to less than $2 \times 10^{-9}$ Torr, was equipped with an EA-200MCD hemispherical electron analyzer with a dual X-ray source using AlK$\alpha$($h\nu$ = 1486.6 eV) at 120 W, at 30 mA, with C (1s) as energy reference (284.6 eV). The sample was prepared in a on 4 mm × 4 mm pellets 0.5 mm thick, and outgassing to a pressure below about $2 \times 10^{-8}$ Torr at 150 °C in the instrument pre-chamber. XPS experiments were carried out at the Central Service for Research Support (SCAI) of the University of Córdoba.

In an Autochem 2920 analyser (Micromeritics Instrument Corp., Norcross, GA, USA) was carried out the Temperature-programmed reduction (TPR) measurements, 200 mg of the solids was used to carried out the experiment with a flow of 40 mL·min$^{-1}$ of a 5% $H_2$/Ar stream. The temperature was ramped from room temperature to 500 °C at 10 °C·min$^{-1}$.

### 2.3. Photocatalytic Experiments

Photocatalytic experiments were performed in a 30 mL double -mouthed heart-shaped reactor under UV light irradiation (UV Spotlight source Lightningcure™ L8022, Hamamatsu, maximum emission at 365 nm) or solar irradiation (Newport, Xe lamp). Light was focalized on the sample compartment through an optic fiber. In a typical process, 5 mg of catalyst was dispersed into 5 mL of glycerol/water (10% *v*/*v*) solution. Reactions were performed under an inert atmosphere, achieved by bubbling a nitrogen flow (20 mL·min$^{-1}$) for 30 min. The catalyst suspension was continuously stirred (800 rpm) and the reactor was thermostated at 20 °C. A picture of the photocatalytic reactor is shown in Figure 2.

**Figure 2.** Picture of the photocatalytic reactor used in the hydrogen production from glycerol photo-reforming.

Hydrogen was analyzed by sampling with a pressure-lock precision analytical syringe (Valco VICI Precision Syringes, 1 mL, leak-tight to 250 psi) from the head space after 3 and 6 h of irradiation. Analyses were performed on an Agilent Technologies 7890A gas chromatograph equipped with a Supelco Carboxen™ 1010 Plot column with TCD detector. The separation was performed at 70 °C for 2 min, followed by heating to 120 °C (ramp of 10 °C·min$^{-1}$), and was left for 8 min (total analysis time, 15 min). All reactions were performed in duplicate, with the standard deviation being below 3%. The calibration plot (Figure S1) and a typical chromatogram for hydrogen quantification (Figure S2) are given in the Supplementary Materials.

## 3. Results and Discussion

The synthesized catalysts were characterized from the structural and chemical point of view with a wide variety of techniques. The chemical composition was determined by ICP-MS and the results, presented in Table 1, evidenced a good incorporation of nickel, with values being quite close to the nominal content (0.5% by weight).

**Table 1.** Metallic content of the solids as determined by inductively-coupled plasma mass spectrometry (ICP-MS). DP: deposition–precipitation; IMP: impregnation.

| Catalyst | Ni Nominal Content (weight%) | Experimental Ni Content (ICP-MS, weight%) |
|---|---|---|
| Ni-0.5-DP | 0.50 | 0.55 |
| Ni-0.5-IMP | 0.50 | 0.47 |

X-ray diffraction patterns were obtained and used to obtain structural information of the catalysts, and the results are presented in Figure S3. The Evonik P25 support clearly shows the diffraction lines corresponding to anatase (80%) and rutile (20%) phases, which are not affected by nickel incorporation, independent of the synthetic method or reduction treatment. In addition, consistent with the small metal loading and the homogeneous nickel dispersion evidenced by TEM micrographs, no signals associated to nickel species are observed.

Temperature-programmed reduction (TPR) measurements were carried out for both Ni-0.5-DP and Ni-0.5-IMP catalysts, and the results are presented in Figure 3 (right). These results showed that the reduction peaks associated with nickel species begin at temperatures below 200 °C but extend to 400 °C.

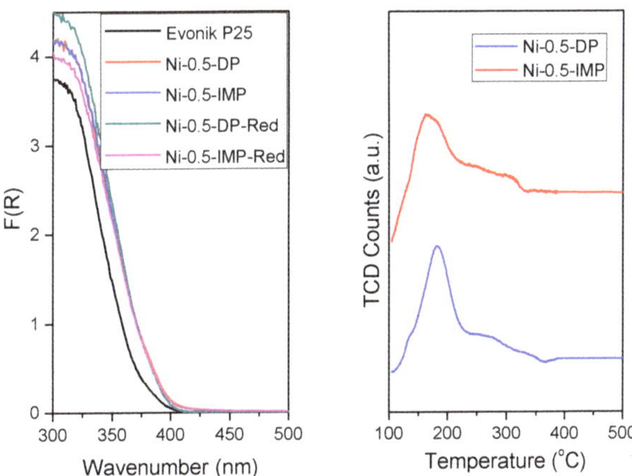


**Figure 3.** UV-vis spectra (**left**) and temperature-programmed reduction profiles (**right**) obtained for the catalysts prepared in this work.

It is well known that NiO is usually the main surface species when nickel is deposited in high loads on metal oxide-type supports. However, certain species from the Ni–support interaction can be observed depending on the physicochemical properties of the support. It has been reported that the reduction of certain nickel species is difficult in supported nickel catalysts, with this difficulty being proportional to the strength of the Ni–support interaction [24]. In general, the Ni–support interaction falls into three categories: i) an absence of interaction, which occurs when the support acts as a mere dispersing agent, ii) a weak interaction, usually associated with the presence of small nickel nanoparticles deposited on the support; and iii) a very strong interaction, involving the formation of a new surface species (creation of new chemical bonds). The degree of interaction depends on the nickel charge (particle size) and the calcination temperature of the catalyst [24].

It has been reported that the reduction of unsupported NiO takes place at temperatures of around 220 °C [24], while the presence of metal–support interactions extend the nickel reduction process to higher temperatures. However, Petrik et al. associated the observed reduction peak at 200 °C to the reduction of $Ni_2O_3$ (ions with formal oxidation state higher than +2) to NiO [25]. In this sense, Carley et al. demonstrated, through XPS studies, the massive formation of surface $Ni^{3+}$ species after the calcination of the solid at temperatures above 300 °C [26]. Finally, the reduction peaks observed at higher temperatures (300–600 °C) were associated with the reduction of the previously formed NiO species or to the reduction of small nanoparticles interacting with the $TiO_2$ support [25].

Based on the above considerations, the observed reduction peak at about 200 °C could be associated with either the reduction of bulk NiO or with the reduction of $Ni^{+3}$ species (nickel ions with a formal oxidation state higher than +2) [25,26]. Given the low nickel loading (0.5%) as well as the small particle sizes reported by TEM (2 and 4 nm for Ni-0.5-DP and Ni-0.5-IMP, respectively), it is more feasible to associate the reduction peak at 200 °C with the reduction of $Ni^{+3}$ species present in the catalyst. Reduction peaks observed at higher temperatures would be associated with small NiO particles interacting with the titania support. According to these results, the temperature chosen for catalyst reduction was set to 400 °C.

Band-gap energy values of the semiconductors were determined from UV-Vis spectra. The method for the determination of band gap values is shown in Figure S4 using the example of Ni-0.5-IMP-Red. As can be seen, the modification of the reference titania material (Evonik P25) by nickel incorporation resulted in a slight decrease in the band gap (Table 2), with the absorption being shifted to the visible spectrum (Figure 3, left)

**Table 2.** Band-gap energy values of the solids as determined by UV-Vis spectroscopy.

| Catalyst | Eg (eV) | Catalyst | Eg (eV) |
|---|---|---|---|
| P25 | 3.11 | | |
| Ni-0.5-DP | 3.02 | Ni-0.5-DP-Red | 3.07 |
| Ni-0.5-IMP | 3.00 | Ni-0.5-IMP-Red | 3.01 |

TEM micrographs of the different solids are shown in Figures 4–6, and the particle size distribution is shown in Figure S5. Ni particle sizes were determined using ImageJ software. The deposition–precipitation method resulted in particles with an average size of 2 nm, whereas the impregnation method led to more heterogeneously-distributed sizes (Figure S5), with the average particle size being 4–5 nm. Particle sizes did not vary significantly after the pre-reduction treatment.

Furthermore, the Ni particle size did not vary significantly after the first use. In the case of the utilization of UV light, there were no changes either after the second use. On the contrary, when solar irradiation was applied, the Ni particle size in the Ni-0.5-DP sample increased up to 5 nm (Figure 6).

The surface chemical composition of the solids was studied by XPS, and the main results are summarized in Table 3. As far as the Ti (2p3/2) signal is concerned, there were no significant changes after the incorporation of nickel, with the signal appearing at ca. 458.5 eV, which is a typical value for $Ti^{4+}$ in $TiO_2$. Regarding the Ni 2p3/2 signal, it has been reported that binding energies at 852.6, 854.6, and 856.1 eV correspond to $Ni^0$, $Ni^{+2}$, and $Ni^{+3}$, respectively [25,27]. As we have commented previously, based on XPS data, Carley et al. unequivocally demonstrated the existence of $Ni^{+3}$ species in nanostructured solids calcined at temperatures above 300 °C [26].

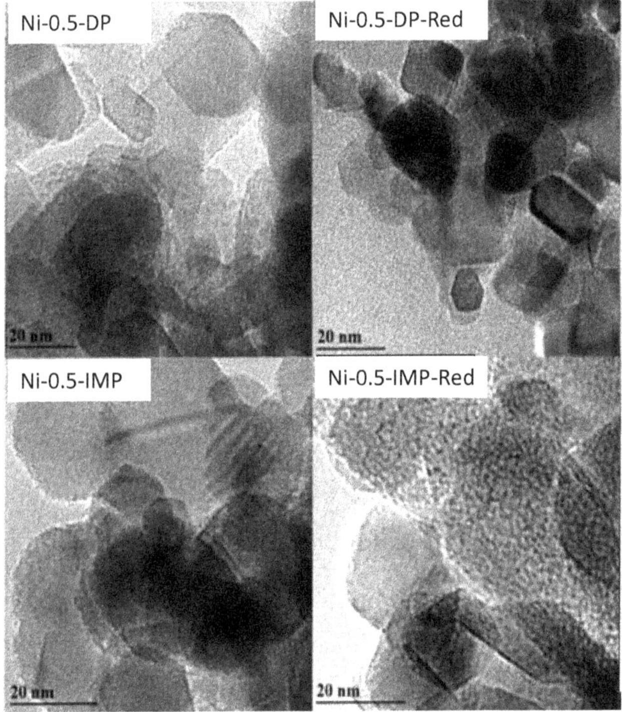


**Figure 4.** TEM micrographs of the fresh (unused) Ni catalysts.

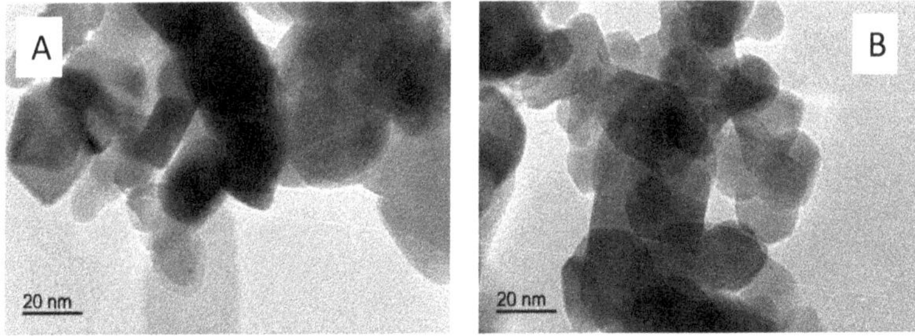


**Figure 5.** TEM micrographs of Ni-0.5-DP solid after the reaction under UV (**A**) or solar (**B**) irradiation.

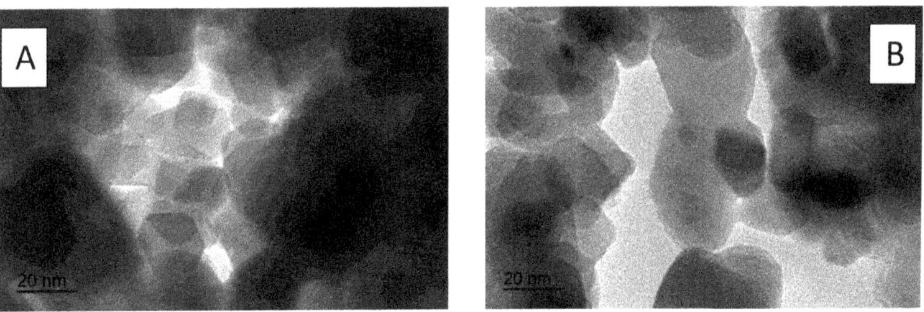


**Figure 6.** TEM micrographs of Ni-0.5-DP solid after the second reutilization using UV (**A**) or solar (**B**) irradiation.

**Table 3.** Ni (2p3/2), Ti (2p3/2) and Cl (2p) binding energies (eV) as determined by XPS.

| Catalyst | Ni (2p3/2) | Ti (2p3/2) | Cl (2p) |
|---|---|---|---|
| Ni-0.5-DP | 855.9 | 458.8 | 199.9 |
| Ni-0.5-DP-Red | 855.0 | 458.7 | - |
| Ni-0.5-IMP | 856.1 | 458.8 | 200.0 |
| Ni-0.5-IMP-Red | 855.1 | 458.5 | - |
| Ni-0.5-DP 1 using UV | 854.9 | 458.6 | - |
| Ni-0.5-DP 1 using solar | 855.1 | 458.6 | - |
| Ni-0.5-DP 2 using UV | 855.2 | 458.4 | - |
| Ni-0.5-DP 2 using solar | 854.9 | 458.4 | - |

Moreover, Petrik et al., working with nanosized nickel oxides, found that binding energies at 855.3 and 856.7 eV were typical for Ni/$TiO_2$ systems, with the signal at around 856 eV being preferably associated with $Ni_2O_3$ rather than with NiO, whose signal appeared at around 855 eV. The authors thus speculated the existence of Ti–Ni–O interactions that were reflected as $Ni^{+3}$ species in the XPS spectra [25].

In this work, the XPS data associated with Ni (2p3/2) signals are presented in Table 3 and Figure S6. The spectra showed a signal at ca. 856 eV for fresh unreduced solids which was assigned to $Ni^{+3}$ species, whereas, after the reduction treatment, the signal shifts to 855 eV as a result of the reduction of $Ni^{+3}$ to $Ni^{+2}$ species. This is in agreement with the reduction peak observed in the TPR profile at around 200 °C. Moreover, in the XPS analysis of the non-reduced catalysts used in a photo-reforming process (both UV and solar), the Ni (2p3/2) signal appears at ca. 855 eV, indicating that during the photocatalytic process, the in-situ reduction of $Ni^{+3}$ to $Ni^{+2}$ species takes place (Figure S6). No signal associated to Ni metal was detected in XPS profiles, even for the reduced solids, and so the catalyst reduction at 400 °C would not be strong enough to carry out the NiO reduction to $Ni^0$, or else the hypothetically formed $Ni^0$ would re-oxidize in contact with air. In this sense, Ju et al. have already reported the absence of the Ni (0) peak at 852 eV after the reductive treatment of nickel-containing absorbents, were was associated with the difficulty of reducing NiO to metallic nickel [28].

It is also interesting to note that XPS revealed the presence of chlorine from the precursor in fresh unreduced solids (0.56 and 0.80 atomic % for Ni-0.5-IMP and Ni-0.5-DP, respectively). Such chlorine atoms were eliminated either during pre-reduction treatment as HCl or during the photocatalytic reaction.

$H_2$ production from glycerol photo-reforming on fresh unreduced catalysts after 3 and 6 h of UV (A) or solar (B) irradiation are given in Figure 7. For the sake of comparison, results obtained for the reference material (Evonik P25) have also been included.

A first conclusion from Figure 7 is that hydrogen production using UV light is always higher than that achieved with solar light. This is hardly surprising, considering that the former irradiation source is more energetic. Moreover, Ni incorporation to $TiO_2$ (irrespective of the method) led to an increase in hydrogen production. For instance, when UV light was used, hydrogen production increased from 166 micromole·$g^{-1}$ (Evonik P25) up to 534 (Ni-0.5-DP) or 551 (Ni-0.5-IMP) after $t = 6$ h. Such an increase is even more significant when visible light was used. The observed shift of UV-Vis absorption to the visible region on the introduction of Ni could account for this effect.

A comparison of hydrogen production on pre-reduced (Figure 8) and untreated systems (Figure 7) allows us to conclude that catalyst pre-reduction treatment significantly increases catalytic activity (3–5 fold or 8–9 fold for experiments under UV or solar irradiation, respectively). As with the untreated systems, there are no large differences in their catalytic behavior depending on the synthesis procedure (DP or IMP). In the pre-reduced systems, those synthesized by DP have 39% greater activity, which could be due to the more homogeneous particle size distribution of Ni.

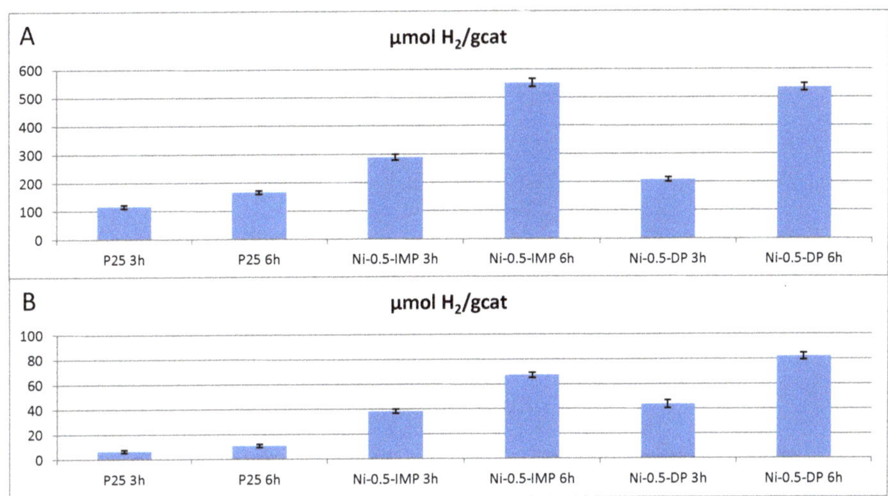


**Figure 7.** $H_2$ production from glycerol photo-reforming on fresh, unreduced solids using UV (**A**) and solar (**B**) irradiation.

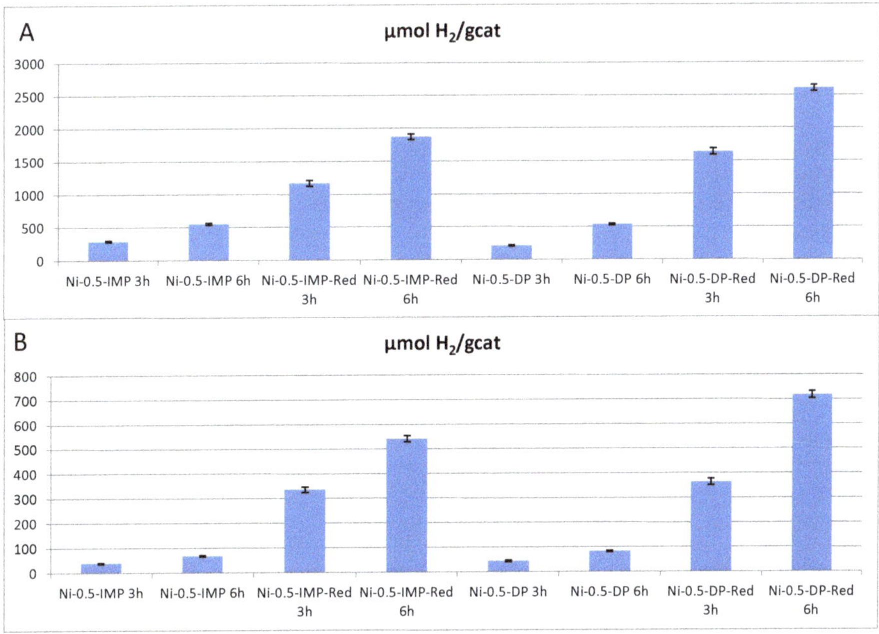


**Figure 8.** $H_2$ production via glycerol photo-reforming on untreated and pre-reduced solids using UV (**A**) and solar (**B**) irradiation.

The highest hydrogen production values (2606 $H_2$ micromole·$g^{-1}$) corresponded to Ni-0.5-DP-Red for $t = 6$ h. This value is similar to that achieved in previous studies on 0.2% Pt [29], which is quite promising considering that Pt is ca. 2000 times more expensive than Ni.

Some authors, such as Bahruji et al. [30], have described the influence of the metal oxidation state on the photocatalytic process. As can be seen in Figure 9B, the electron transfer from titania to NiO is thermodinamically impeded. On the contrary, the pre-reduction of the solid (Figure 9A) results in electron transfer from titania to Ni(0) being favored, with the metal thus acting as an electron trap and preventing electron–hole recombination.

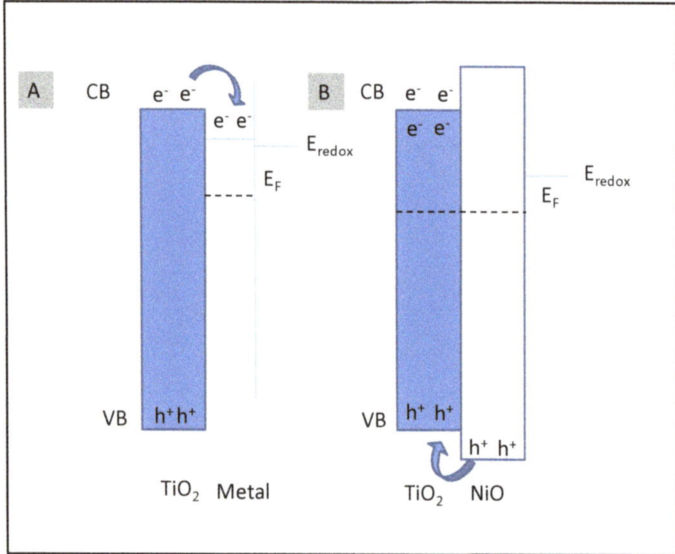


**Figure 9.** Energy levels of (**A**) $TiO_2$/metal and (**B**) $TiO_2$/NiO, adapted from Bahruji et al [30].

Furthermore, Caravaca et al. [31], studying hydrogen photo-production from sugars on Ni-based catalysts, observed an induction period (in their case, 60 min) required for in-situ reduction of NiO to Ni. After that period of time, the hydrogen production rate of both the untreated and pre-reduced solid was the same. According to these reports, we asume the in-situ reduction of our catalysts during reactions, and so the electron transfer from titania to Ni(0) is favored [30].

Another possible explanation for the observed better catalytic performance of pre-reduced solids as compared to untreated systems is the presence in the latter solids of surface chlorides (a well-known poison for metals arising from the precursor and evidenced by XPS analyses). As mentioned above, those chloride species were not observed in pre-reduced systems as they were eliminated as HCl during hydrogen pretreatment.

In order to cast further light on the effect of Ni oxidation states and the presence of chloride species on catalytic performance, some reutilisation studies were carried out on the Ni-0.5-DP catalyst both under UV and solar irradiation. Therefore, after 6 h irradiation, the solid was recovered by filtration, washed with methanol and acetone and dried at 110 °C. The catalyst was labelled as Ni-0.5-DP 1 using UV and Ni-0.5-DP 1 using solar, depending on the irradiation source. The solids were tested in another reaction, and the catalytic results are shown in Figure 10. As can be seen, hydrogen production dropped from 503 to 212 micromole per gram of catalyst after 6 h of UV irradiation, whereas no significant deactivation was observed under visible light. In any case, catalytic results were far below those achieved with fresh, pre-reduced catalysts.

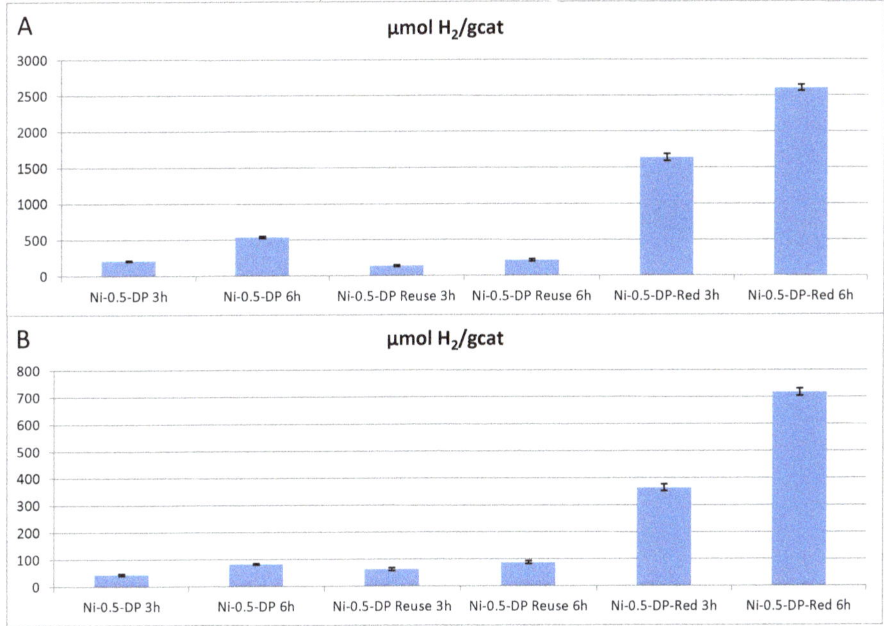


**Figure 10.** Comparison of hydrogen photocatalytic production on untreated Ni-0.5-DP after the first and second use using UV light (**A**) or solar light (**B**). For the sake of comparison, results obtained for the pre-reduced solid (Ni-0.5-DP-Red) have also been included.

TEM micrographs (Figure 5) did not evidence any significant increase in Ni particle size with the first use. After the second use, the metal particle size only slightly increased for solar irradiation studies.

XPS experiments (Table 3) showed that surface chlorides had already been eliminated after the first use, and it is assumed that there had been an in-situ reduction of nickel species. Thus, neither the presence of chlorides nor the in-situ reduction of NiO can account for the possitive effect of pre-reduction treatment at 400 °C on catalytic performance. It is possible that such a pre-treatment induced a strong metal–support interaction (SMSI) [32,33] which somehow favored the subsequent catalytic performance. This SMSI would be favored on homogeneously-distributed particles achieved by the deposition–precipitation method, which could explain the above-mentioned better catalytic performance of Ni-0.5-DP-Red as compared to Ni-0.5-IMP-Red. Nevertheless, these hypothetical Ni–support interactions were not detected by XPS measurements and therefore require further studies.

## 4. Conclusions

Hydrogen production through the glycerol photo-reforming of titania-based systems was significantly improved on the incorporation of nickel (0.5% by weight) through deposition–precipitation (DP) or impregnation (IMP) methods. This improvement was more pronounced when solar light was used as the irradiation source. The absorption shift to the visible range in the presence of Ni (evidenced by UV-Vis) could account for this effect. The pre-reduction of the systems prior to the catalytic essays led to a substantial improvement in catalytic performance, despite the fact that XPS studies showed that i) nickel species were "in-situ" reduced under working conditions, and ii) surface chloride species (arising from the used precursor, $NiCl_2$) are removed as the reaction proceeds. It is possible that pre-reduction treatment at 400 °C induced a strong metal–support interaction (SMSI) which could be positive to the catalytic performance, although this requires further studies. This SMSI effect would be favored for systems with smaller, more uniformly-distributed Ni particle sizes synthesized by the DP method as compared to the IMP method. This would be consistent with the observed higher activities of Ni-0.5-DP Red as compared to Ni-0.5-IMP Red. In summary, the addition of a small percentage (0.5% by weight) of a transition metal such as Ni (ca. 2000 times cheaper than Pt) resulted in a 15.5-fold increase in the catalytic activity of Evonik P25, producing 2.6 mmol $H_2 \cdot g^{-1}$ after 6 h of UV irradiation. Thus, Ni proved to be a promising metal for use in photo-reforming processes of biomass-derived oxygenated compounds.

**Supplementary Materials:** The following are available online at http://www.mdpi.com/1996-1073/12/17/3351/s1. Figure S1. Hydrogen calibration plot used in this work. Figure S2. A typical chromatogram obtained in the hydrogen quantification process through GC. Figure S3. X-ray diffractograms of Evonik P25, Ni-0.5-DP. Ni-0.5-IMP, Ni-0.5-DP-Red and Ni-0.5-IMP-Red. Figure S4. Band-gap energy (Eg) calculation for the Ni-0.5-IMP-Red catalysts. Figure S5. Particle size distribution of each catalyst. Figure S6. XPS profiles for the Ni 2p3/2 component of each catalyst.

**Author Contributions:** Conceptualization, A.M. and F.J.U.; methodology, J.H-C. and A.M.; validation, F.J.U., A.M. and J.H.-C.; formal analysis, A.M. and J.H-C; investigation, J.M-G., J.M. and J.C.E.; data curation, A.M., J.H.-C.; writing—original draft preparation, J.M.-G. and J.H.C.; writing—review and editing, F.J.U. and A.M.; supervision, J.H-C., F.J.U. and A.M.

**Funding:** The authors are thankful to MINECO-ENE2016-81013-R (AEI/FEDER, EU).

**Acknowledgments:** The scientific support from the Central Service for Research Support (SCAI) at the University of Cordoba is acknowledged.

**Conflicts of Interest:** The authors declare no conflict of interest.

## References

1. Melián, E.P.; López, C.R.; Santiago, D.E.; Quesada-Cabrera, R.; Méndez, J.A.O.; Rodríguez, J.M.D.; Díaz, O.G. Study of the photocatalytic activity of Pt-modified commercial $TiO_2$ for hydrogen production in the presence of common organic sacrificial agents. *Appl. Catal. A Gen.* **2016**, *518*, 189–197. [CrossRef]
2. Granada Ramírez, O.; Gutierrez Arzaluz, M.; Ávila Jiménez, M.; Fernández Sánchez, L.; Aguilar Pliego, J.; Mugica Álvarez, V.; Torres Rodríguez, M. Oxidación de glicerol en fase acuosa con platino soportado en óxido de titanio. *Superf. Vacío* **2016**, *29*, 9–13.
3. Avgouropoulos, G.; Ioannides, T.; Kallitsis, J.K.; Neophytides, S. Development of an internal reforming alcohol fuel cell: Concept, challenges and opportunities. *Chem. Eng. J.* **2011**, *176*, 95–101. [CrossRef]
4. Du, C.; Mo, J.; Li, H. Renewable Hydrogen Production by Alcohols Reforming Using Plasma and Plasma-Catalytic Technologies: Challenges and Opportunities. *Chem. Rev.* **2015**, *115*, 1503–1542. [CrossRef] [PubMed]
5. Singh, R.; White, D.; Demirel, Y.; Kelly, R.; Noll, K.; Blum, P. Uncoupling Fermentative Synthesis of Molecular Hydrogen from Biomass Formation in Thermotoga maritima. *Appl. Environ. Microbiol.* **2018**, *84*, e00998-18. [CrossRef] [PubMed]
6. Singh, R.; Tevatia, R.; White, D.; Demirel, Y.; Blum, P. Comparative kinetic modeling of growth and molecular hydrogen overproduction by engineered strains of Thermotoga maritima. *Int. J. Hydrogen Energy* **2019**, *44*, 7125–7136. [CrossRef]
7. Bednarczyk, K.; Stelmachowski, M.; Gmurek, M. The Influence of Process Parameters on Photocatalytic Hydrogen Production. *Environ. Prog. Sustain. Energy* **2018**, *38*, 680–687. [CrossRef]
8. Christoforidis, K.C.; Fornasiero, P. Photocatalytic Hydrogen Production: A Rift into the Future Energy Supply. *ChemCatChem* **2017**, *9*, 1523–1544. [CrossRef]
9. Wang, C.; Cai, X.; Chen, Y.; Cheng, Z.; Luo, X.; Mo, S.; Jia, L.; Lin, P.; Yang, Z. Improved hydrogen production from glycerol photoreforming over sol-gel derived $TiO_2$ coupled with metal oxides. *Chem. Eng. J.* **2017**, *317*, 522–532. [CrossRef]
10. Martínez, F.M.; Albiter, E.; Alfaro, S.; Luna, A.L.; Colbeau-Justin, C.; Barrera-Andrade, J.M.; Remita, H.; Valenzuela, M.A. Hydrogen Production from Glycerol Photoreforming on $TiO_2$/HKUST-1 Composites: Effect of Preparation Method. *Catalysts* **2019**, *9*, 338. [CrossRef]
11. Barreca, D.; Carraro, G.; Gombac, V.; Gasparotto, A.; Maccato, C.; Fornasiero, P.; Tondello, E. Supported metal oxide nanosystems for hydrogen photogeneration: Quo vadis? *Adv. Funct. Mater.* **2011**, *21*, 2611–2623. [CrossRef]

12. Chen, X.; Selloni, A. Introduction: Titanium Dioxide ($TiO_2$) Nanomaterials. *Chem. Rev.* **2014**, *114*, 9281–9282. [CrossRef] [PubMed]
13. Kumaravel, V.; Mathew, S.; Bartlett, J.; Pillai, S.C. Photocatalytic hydrogen production using metal doped $TiO_2$: A review of recent advances. *Appl. Catal. B Environ.* **2019**, *244*, 1021–1064. [CrossRef]
14. Daghrir, R.; Drogui, P.; Robert, D. Modified $TiO_2$ For Environmental Photocatalytic Applications: A Review. *Ind. Eng. Chem. Res.* **2013**, *52*, 3581–3599. [CrossRef]
15. Zhang, J.; Huang, Y.; Dan, Y.; Jiang, L. P3HT/Ag/$TiO_2$ ternary photocatalyst with significantly enhanced activity under both visible light and ultraviolet irradiation. *Appl. Surf. Sci.* **2019**, *488*, 228–236. [CrossRef]
16. Jovic, V.; Chen, W.-T.; Sun-Waterhouse, D.; Blackford, M.G.; Idriss, H.; Waterhouse, G.I.N. Effect of gold loading and $TiO_2$ support composition on the activity of Au/$TiO_2$ photocatalysts for $H_2$ production from ethanol–water mixtures. *J. Catal.* **2013**, *305*, 307–317. [CrossRef]
17. López-Tenllado, F.J.; Hidalgo-Carrillo, J.; Montes, V.; Marinas, A.; Urbano, F.J.; Marinas, J.M.; Ilieva, L.; Tabakova, T.; Reid, F. A comparative study of hydrogen photocatalytic production from glycerol and propan-2-ol on M/$TiO_2$ systems (M = Au, Pt, Pd). *Catal. Today* **2017**, *280*, 58–64. [CrossRef]
18. Bowker, M.; Morton, C.; Kennedy, J.; Bahruji, H.; Greves, J.; Jones, W.; Davies, P.R.; Brookes, C.; Wells, P.P.; Dimitratos, N. Hydrogen production by photoreforming of biofuels using Au, Pd and Au–Pd/$TiO_2$ photocatalysts. *J. Catal.* **2014**, *310*, 10–15. [CrossRef]
19. Carraro, G.; MacCato, C.; Gasparotto, A.; Montini, T.; Turner, S.; Lebedev, O.I.; Gombac, V.; Adami, G.; Van Tendeloo, G.; Barreca, D.; et al. Enhanced hydrogen production by photoreforming of renewable oxygenates through nanostructured $Fe_2O_3$ polymorphs. *Adv. Funct. Mater.* **2014**, *24*, 372–378. [CrossRef]
20. Iriondo, A.; Barrio, V.L.; Cambra, J.F.; Arias, P.L.; Guemez, M.B.; Sanchez-Sanchez, M.C.; Navarro, R.M.; Fierro, J.L.G. Glycerol steam reforming over Ni catalysts supported on ceria and ceria-promoted alumina. *Int. J. Hydrogen Energy* **2010**, *35*, 11622–11633. [CrossRef]
21. Chen, W.-T.; Chan, A.; Sun-Waterhouse, D.; Llorca, J.; Idriss, H.; Waterhouse, G.I.N. Performance comparison of Ni/$TiO_2$ and Au/$TiO_2$ photocatalysts for $H_2$ production in different alcohol-water mixtures. *J. Catal.* **2018**, *367*, 27–42. [CrossRef]
22. Clarizia, L.; Vitiello, G.; Pallotti, D.K.; Silvestri, B.; Nadagouda, M.; Lettieri, S.; Luciani, G.; Andreozzi, R.; Maddalena, P.; Marotta, R. Effect of surface properties of copper-modified commercial titanium dioxide photocatalysts on hydrogen production through photoreforming of alcohols. *Int. J. Hydrogen Energy* **2017**, *42*, 28349–28362. [CrossRef]
23. Luciani, G.; Vitiello, G.; Clarizia, L.; Abdelraheem, W.; Esposito, S.; Bonelli, B.; Ditaranto, N.; Vergara, A.; Nadagouda, M.; Dionysiou, D.D.; et al. Near UV-irradiation of CuOx-impregnated $TiO_2$ providing active species for $H_2$ production through methanol photoreforming. *ChemCatChem* **2019**, *45324*, 1–14.
24. Li, C.; Chen, Y.W. Temperature-programmed-reduction studies of nickel oxide/alumina catalysts: Effects of the preparation method. *Thermochim. Acta* **1995**, *256*, 457–465. [CrossRef]
25. Petrik, I.S.; Krylova, G.V.; Lutsenko, L.V.; Smirnova, N.P.; Oleksenko, L.P. XPS and TPR study of sol-gel derived M/$TiO_2$ powders (M = Co, Cu, Mn, Ni). *Chem. Phys. Technol. Surf.* **2015**, *6*, 179–189.
26. Carley, A.F.; Jackson, S.D.; O'Shea, J.N.; Roberts, M.W. The formation and characterisation of Ni3+—An X-ray photoelectron spectroscopic investigation of potassium-doped Ni(1 1 0)-O. *Surf. Sci.* **1999**, *440*, L868–L874. [CrossRef]
27. Janlamool, J.; Jongsomjit, B. Characteristics and catalytic properties of Ni/Ti-Si composite oxide catalysts via $CO_2$ hydrogenation. *Eng. J.* **2017**, *21*, 45–55.
28. Ju, F.; Wang, M.; Wu, T.; Ling, H. The role of NiO in reactive adsorption desulfurization over NiO/ZnO-$Al_2O_3$-$SiO_2$ adsorbent. *Catalysts* **2019**, *9*, 79. [CrossRef]
29. Morales, J. Hydrogen Production by Glycerol Photo-Reforming, Using Noble and Transition Metals Supported on Titanium Dioxide. Master's Thesis, University of Córdoba (SPAIN), Cordoba, Spain, 2018.
30. Bahruji, H.; Bowker, M.; Davies, P.R.; Kennedy, J.; Morgan, D.J. The importance of metal reducibility for the photo-reforming of methanol on transition metal-$TiO_2$ photocatalysts and the use of non-precious metals. *Int. J. Hydrogen Energy* **2015**, *40*, 1465–1471. [CrossRef]
31. Caravaca, A.; Jones, W.; Hardacre, C.; Bowker, M. $H_2$ production by the photocatalytic reforming of cellulose and raw biomass using Ni, Pd, Pt and Au on titania. *Proc. R. Soc. A Math. Phys. Eng. Sci.* **2016**, *472*, 20160054. [CrossRef]

32. Colmenares, J.C.; Magdziarz, A.; Aramendia, M.A.; Marinas, A.; Marinas, J.M.; Urbano, F.J.; Navio, J.A. Influence of the strong metal support interaction effect (SMSI) of Pt/$TiO_2$ and Pd/$TiO_2$ systems in the photocatalytic biohydrogen production from glucose solution. *Catal. Commun.* **2011**, *16*, 1–6. [CrossRef]
33. Szabó, F.Z.G. *Solymosi in "Proceedings, 2nd International Congress on Catalysis"*; Technip: Paris, France, 1961; p. 1627.

*Article*

# $Fe_3O_4$-PDA-Lipase as Surface Functionalized Nano Biocatalyst for the Production of Biodiesel Using Waste Cooking Oil as Feedstock: Characterization and Process Optimization

**Tooba Touqeer [1], Muhammad Waseem Mumtaz [1,*], Hamid Mukhtar [2], Ahmad Irfan [3,4], Sadia Akram [1], Aroosh Shabbir [2], Umer Rashid [5,*], Imededdine Arbi Nehdi [6,7] and Thomas Shean Yaw Choong [8]**

1 Department of Chemistry, University of Gujrat, Gujrat 50700, Pakistan; 17101707-035@uog.edu.pk (T.T.); sadia.nano@yahoo.com (S.A.)
2 Institute of Industrial Biotechnology, Government College University, Lahore 54000, Pakistan; hamidwaseer@yahoo.com (H.M.); aroosh-gcu@hotmail.com (A.S.)
3 Research Center for Advanced Materials Science, King Khalid University, Abha 61413, P.O. Box 9004, Saudi Arabia; irfaahmad@gmail.com
4 Department of Chemistry, Faculty of Science, King Khalid University, Abha 61413, P.O. Box 9004, Saudi Arabia
5 Institute of Advanced Technology, Universiti Putra Malaysia, Serdang 43400 UPM, Selangor, Malaysia
6 Chemistry Department, College of Science, King Saud University, Riyadh 11451, Saudi Arabia; inahdi@ksu.edu.sa
7 Laboratoire de Recherche LR18ES08, Chemistry Department, Science College, Tunis El Manar University, Tunis 2092, Tunisia
8 Department of Chemical and Environmental Engineering, Universiti Putra Malaysia, Serdang 43400 UPM, Selangor, Malaysia; csthomas@upm.edu.my
* Correspondence: muhammad.waseem@uog.edu.pk (M.W.M.); umer.rashid@upm.edu.my (U.R.); Tel.: +60-3-9769-7393 (U.R.)

Received: 16 October 2019; Accepted: 3 November 2019; Published: 31 December 2019

**Abstract:** Synthesis of surface modified/multi-functional nanoparticles has become a vital research area of material science. In the present work, iron oxide ($Fe_3O_4$) nanoparticles prepared by solvo-thermal method were functionalized by polydopamine. The catechol groups of polydopamine at the surface of nanoparticles provided the sites for the attachment of *Aspergillus terreus* AH-F2 lipase through adsorption, Schiff base and Michael addition mechanisms. The strategy was revealed to be facile and efficacious, as lipase immobilized on magnetic nanoparticles grant the edge of ease in recovery with utilizing external magnet and reusability of lipase. Maximum activity of free lipase was estimated to be 18.32 U/mg/min while activity of $Fe_3O_4$-PDA-Lipase was 17.82 U/mg/min (showing 97.27% residual activity). The lipase immobilized on polydopamine coated iron oxide ($Fe_3O_4$_PDA_Lipase) revealed better adoptability towards higher levels of temperature/pH comparative to free lipase. The synthesized ($Fe_3O_4$_PDA_Lipase) catalyst was employed for the preparation of biodiesel from waste cooking oil by enzymatic transesterification. Five factors response surface methodology was adopted for optimizing reaction conditions. The highest yield of biodiesel (92%) was achieved at 10% $Fe_3O_4$_PDA_Lipase percentage concentration, 6:1 $CH_3OH$ to oil ratio, 37 °C temperature, 0.6% water content and 30 h of reaction time. The $Fe_3O_4$-PDA-Lipase activity was not very affected after first four cycles and retained 25.79% of its initial activity after seven cycles. The nanoparticles were characterized by FTIR (Fourier transfer infrared) Spectroscopy, XRD (X-ray diffraction) and TEM (transmission electron microscopy), grafting of polydopamine on nanoparticles was confirmed by FTIR and formation of biodiesel was evaluated by FTIR and GC-MS (gas chromatography-mass spectrometry) analysis.

**Keywords:** biodiesel; transesterification; *Aspergillus terreus* lipase; polydopamine; immobilization; RSM; fuel properties

## 1. Introduction

Overpopulation, urbanization, industrialization and augmentation of personal transport have increased the petroleum consumption immensely. It is predicted that petroleum reserves will become depleted soon if the rate of its consumption continue to increase with the same rate [1]. Depletion of non-renewable fuel reserves, elevation in price of fossil fuels and pollution caused by usage of petroleum products has spurred the production of eco-friendly, inexpensive alternative energy source that can reduce petroleum consumption and that is biodiesel [2]. Biodiesel has emerged as an alternative of petroleum due to its renewable nature, biodegradability, and interest of consumers in nature friendly products [3].

Feedstock selection for biodiesel production is of chief concern because of high prices associated with various feedstock. Edible and non-edible feedstocks have been used by many researchers for the biodiesel synthesis, edible oils including; sunflower, corn, rapeseed and soybean oil, while non edible oils including; Jatropha, *Eruca sativa*, Caster, Jojoba and other oils. Utilizing edible oils as feedstock will cause conflict with food production and food price so it is more desirable to use non edible oil [4]. A very economical feedstock is used cooking oil. Excess of waste frying/cooking oil is produced every day in restaurants and fast food shops. The use of edible oil for excessive frying/cooking at high temperature destroys the structure of triglyceride and produce the free fatty acids in oil, changing its pH, which is harmful for human health [5,6]. The oil is therefore being dumped off through drainage, which causes water pollution. This oil is useless and causes hazards to aquatic life. Using this waste oil to make something valuable is highly appreciated, because it is almost free feedstock for biodiesel production [7,8].

Transesterification is the preferred method for biodiesel production as compared to other techniques [9]. Various strategies of transesterification have been adopted to achieve better yield, purity and fast reaction rates. Catalytic transesterification is frequently used as, the presence of catalyst enhances the solubility of oil in alcohol and increases the reaction rate [10]. Catalysts of transesterification can be homogeneous or heterogeneous. Homogenous catalysts include; acid and alkaline catalyst. Alkaline catalysts are commonly used because of their low cost and higher catalytic activity at ambient conditions, while alkaline catalysts do not work efficiently under higher FFA conditions [11]. Besides that, there are many drawbacks of using these catalysts in excess which includes; formation of toxic waste that need to be neutralized by several washings, hence it causes environmental pollution, contaminated glycerol is produced, purification of this glycerol increases the production cost, partial saponification can lead to the production of soap which makes it difficult to separate glycerol and alkyl esters hence decreases the yield of biodiesel [12]. Furthermore, another major drawback is that these catalysts cannot be reused or recycled.

Heterogenous catalysts include; enzymes (lipase), alkaline earth metals, zirconias of potassium and silicates of titanium etc. The enzyme used for transesterification is usually lipase. Lipase has several advantages over convention alkali catalysts; e.g., there is no need of purification of biodiesel after transesterification thus no toxic waste is produced—especially in case of waste cooking oil, where there are lot of free fatty acids, lipase reduces the chance of saponification [13]. However, there are some reasons that hampered its utilization, which include; high cost, difficulty in its recovery and instability at high temperature and pH. The best strategy to tackle these problems is the immobilization of enzyme on some support. Previously, lipases have been immobilized on several surfaces such as; ceramics, calcium alginate beads and other inorganic matrixes, which increase the thermal stability of the enzyme. However, the activity of the enzyme is revealed to be reduced due to a decrease in conformational flexibility and capability of adsorption on support surface [14]. Another effective way of using bio-catalyst is to immobilize it on nano-support. Different techniques such as; entrapment, adsorption and covalent immobilization are being used for this purpose. Enzymes immobilized on nanoparticles provide the advantage of greater enzymatic activity, better selectivity along with thermal stability, adaptability towards wider pH range, easy recovery and purification [15]. If the nanoparticles used for immobilization are metal oxides, then these particles provide the advantage of low pricing, and higher stability even in harsh conditions, another advantageous attribute of metal oxide nanoparticles (NPs) is their magnetic property. Nanoparticles of iron, nickel, cobalt, chromium, manganese and their oxides show enhanced magnetic moment as compared to other metals [16]. Among these metal oxides, iron oxides—especially magnetite $Fe_3O_4$—shows very strong magnetism and it is also less toxic when compared to nickel and cobalt. Enzyme was first immobilized on magnetic nanoparticles surface by Matsunaga and Kamiya [17]. Immobilizing lipase on magnetite will generate a nano-biocatalyst, which will grant the edge of better activity in harsh conditions and reusability [18].

To reduce the risk of toxicity and make them more biocompatible, the surface of nanoparticles can be modified with different polymers using "graft to and graft from methods". Polymers (natural/synthetic) are used as coatings that also provide active groups at the surface to afford immobilization of lipase on nanoparticles. These polymers mostly contain epoxy and amine groups that react with active groups of lipases. However, the drawback of using polymers with amine group is the need of activation of amine group using some aldehydes, most commonly, amine group activation is done using gutaraldehyde [19]. This activation step increases the cost and preparation time of catalyst.

A versatile coating for the nanoparticles is polydopamine (formed by self-polymerization of dopamine monomers through oxidation in slightly alkaline conditions). Dopamine contains catechol as well as amine group, which efficiently interacts with metal oxide nanoparticles [20]. In an alkaline medium on oxidation, the catechol group of dopamine converts to the indo-5, 6-quinone, which further undergoes series of inter/intra molecular reactions, forming polydopamine grafted on the NPs surface. The residual quinone and catechol groups present on the surface after polymerization are reactive towards nucleophiles such as thiol and amine groups. These groups of lipases covalently immobilize the lipase on polydopamine through Schiff base formation and Michael addition reaction. Lipase immobilized on polydopamine containing nanoparticles show higher efficiency and higher enzyme loading as compared to the immobilization at naked nanoparticles [21]. The present work was therefore planned to prepare magnetic metal oxide nanoparticles by solvothermal method followed by polydopamine grafting and immobilization of *Aspergillus terrus* lipase for development of nano-biocatalyst. The synthesized nano-biocatalyst was then employed for the synthesis of biodiesel.

## 2. Materials and Methods

All the chemicals/reagents used, i.e., $FeCl_3{\cdot}6H_2O$, ethylene glycol, ethylene diamine, sodium acetate, methanol, ethanol, potassium iodide, iodine, n-hexane, chloroform, toluene, acetic acid, distilled water, potassium hydroxide, hydrochloric acid, phenolphthalein and isopropanol etc., were of analytical research grade obtained from Sigma-Aldrich (St. Louis, MO, USA). Lipase was produced from *Aspergillus terreus* AH-F2. The feedstock waste cooking oil was obtained from local restaurant situated in district Gujrat Pakistan.

### 2.1. Preparation of Nanoparticles

$Fe_3O_4$ nanoparticles were prepared/synthesized by solvothermal method by Guoet et al. [22] with some modifications. One gram of $FeCl_3{\cdot}6H_2O$ was added in 20 mL of ethylene glycol and stirred to get a clear solution, then 3 g of sodium acetate and 10 mL of ethylene diamine were added in the solution, which was stirred for 30 min at room temperature. The mixture was then enclosed in a 50 mL Teflon lined autoclave and it was heated in oven for 8 h at 180 °C. The black colour product obtained was washed with successive distilled water/ethanol rinses followed by product drying using a desiccator. Black magnetic powder was obtained after drying.

### 2.2. Grafting of Dopamine on $Fe_3O_4$ Nanoparticles

Dopamine was coated on nanoparticles using "graft from" technique [23]. 0.1 g of $Fe_3O_4$ nanoparticles were suspended in 20 mL of distilled water. Afterwards, 20 mL of 20 milli molar tris-HCl buffer having pH 8.5 was introduced in the suspension. Then, 0.1 g of dopamine hydrochloride was slowly added in the suspension (2 mg/15 s) and the suspension was stirred for 1 h. Nanoparticles coated with polydopamine were formed by the self-polymerization of dopamine in alkaline conditions.

### 2.3. Characterization of Nanoparticles and PDA-Nanoparticles Complex

(a) Nanoparticles were characterized by X-ray diffraction, which helps to identify the crystal phase, dimensions and size of the crystalline material. X'pertpro (PANalyatical) XRD model was used for this purpose. The XRD pattern of $Fe_3O_4$ was attained by use of Cu K alpha radiation of wave length 1.54 Å. The patron was taken between the range 20–80 θ with 0.02 scan step size.

(b) Morphological characterization was carried out by TEM (Tecnai $G^2$ F20 U-TWIN) with an accelerating voltage of 200 kV.

(c) Bare nanoparticles and the nanoparticles after grafting of polydopamine were analysed by FTIR in the scanning range 400–4000 $cm^{-1}$, using the Cary 630 Agilent FTIR spectrometer to check the formation/purity of nanoparticles and to confirm the attachment of dopamine polymers on nanoparticle surface.

### 2.4. Lipase Immobilization on Modified Nanoparticles

Lipase was produced from *Aspergillus terreus* AH-F2 and purified using the method described in previous work of our research group [24]. Into 40 mL of phosphate buffer with pH 7, 0.40 g of lipase was added. The suspension of polydopamine coated magnetic nanoparticles was added slowly to the mixture of lipase with vigorous stirring for 3 h at 4 °C. The nucleophilic groups present in lipase such as thiol and amines reacted with the residual catechole and quinone groups present at the surface of $Fe_3O_4$-PDA nanoparticles by Schiff base and Michael addition mechanisms and were immobilized without the need of any coupling and activating agents (Figure 1). The protein content before and after the immobilization process was determined to check the immobilization yield using Bradford's method. The formed nano-biocatalyst was washed several times with phosphate buffer to remove any unreacted lipase and then freeze dried [23,25].

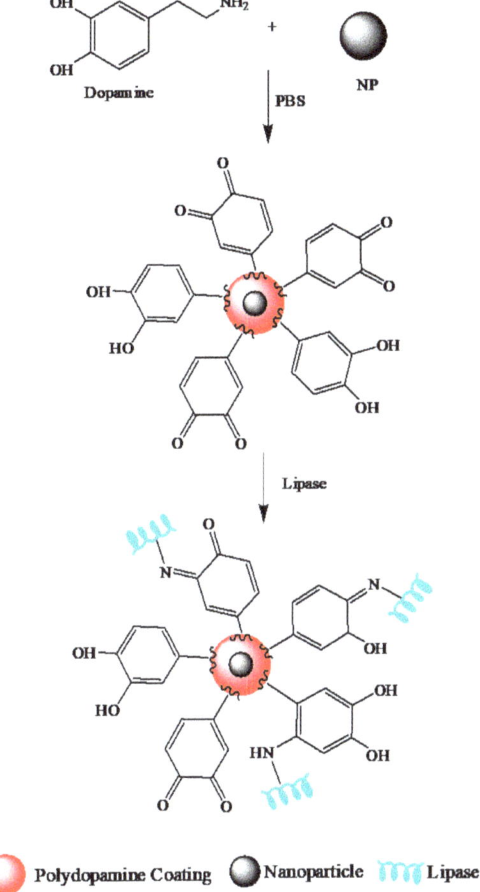


**Figure 1.** Schematic representation of lipase immobilization on dopamine coated.

*2.5. Lipase Activity Assay (Free and Immobilized Form)*

Lipase assay for free and immobilized lipase was performed by using the titrimetric method [24]. The activity of lipase was measured by the titration of fatty acids, which were produced from oil while reacting with enzyme. The assay reaction mixture (consisted of 1 mL of free lipase (produced from *Aspergillus terreus* AH-F2) while 0.1 g of nano-biocatalyst in the case of immobilized lipase and 10 mL of olive oil (10%)) was homogenized in 10 g of gum acacia and phosphate buffer (5 mL) with pH 7 followed by the addition of 0.6% calcium chloride (2 mL). This mixture was then incubated for 1 h in a water bath shaker at 37 °C. After incubation, the enzymatic reaction was ceased by adding 20 mL of ethanol/acetone mixture (1:1) following the addition of few phenolphthalein drops. The quantity of fatty acids released during this period by the action of lipase was titrated by using 0.1 N NaOH solution.

One unit of lipase activity (U) was defined as the amount of enzyme which released one micro mole (µmol) of fatty acid per minute under specified assay conditions.

Lipase units were determinedLipase units were determined as follows:

$$\text{Lipase Activity (U/mg/min)} = \frac{\Delta V \times N}{m(\text{sample})} \times \frac{1000}{60}$$

where $\Delta V = V2 - V1$; V1 = volume of NaOH consumed against control flask; V2 = volume of NaOH consumed against experimental flask; N = normality of NaOH; m(sample) = mass of enzyme extract; 60 = time of incubation (min) for bacterial lipase.

The effects of pH and temperature on the activity of free and immobilized lipase were studied using the above-mentioned assay method. The impact of pH on catalytic activity was investigated by incubating the sample at 37 °C with the phosphate buffer within pH range 5 to 10, whereas the impact of temperature in the range 25–50 °C was also studied at optimum pH levels.

### 2.6. Collection and Characterization of Feedstock

WCO (waste cooking oil) was collected from the local restaurant for the biodiesel production. The feedstock was purified by filtration to remove any food chunks and inorganic material in the waste oil. The physico-chemical properties of WCO biodiesel were analysed to reveal the quality of the used oil. Iodine number, specific gravity, density, peroxide value, acid value, saponification value and refractive index of the oil were calculated using standard analytical AOCS (American oil chemist's society) methods.

### 2.7. Experimental Design

Process of biodiesel production using $Fe_3O_4$_PDA_Lipase was optimized using five factor CCRSM (central composite response surface methodology) provided by Design Expert version 10.0.7 (State-Ease Inc., Minneapolis, MN, USA). The five independent variables chosen for the investigation of the optimization process with their ranges are A (methanol/oil ratio 3:1 to 9:1), B (biocatalyst concentration 1 to 10%), C (reaction temperature 20 to 50 °C), D (reaction time 12 to 48 h) and E (water content 0.2 to 1%). Fifty reactions were carried out as per experimental design.

A reaction mixture containing a specified amount of oil, nano-biocatalyst, methanol and water were allowed to react in a conical flask placed in shaking incubator, while reaction parameters were set in accordance to the RSM (response surface methodology) design. After completion of reaction, the glycerol was separated from biodiesel. Crude biodiesel was purified, and remaining methanol was recovered by using a rotary evaporator under the conditions of reduced pressure. GC-FID, having a highly-polar BPX-70 capillary column (30 m × 0.25 mm), was utilized for the analysis of products. The oven temperature was kept at 100 °C for 30 s, then raised to 250 °C at a rate of 10 °C/min, while the temperature of the detector was set at 270 °C. 500 ppm of sample was dissolved in hexane containing methyl heptadecanoate as an internal standard, and 1 μL of this mixture was injected in column [26,27]. The percentage conversion of biodiesel was calculated by given formula [28]

$$\text{FAME (Biodiesel) \%} = \frac{\sum A_{ME} - A_{IS}}{A_{IS}} \times \frac{C_{IS} \times V}{M} \times 100$$

where $\sum A_{ME}$ denotes the sum of peak areas of FAME's peaks, $A_{IS}$ is the peak area of the internal standard peak (methyl heptadecanoate), $C_{IS}$ is the concentration of the internal standard, V is the volume of the internal standard and M is the mass of biodiesel.

### 2.8. Recovery and Recycling of Nano-Biocatalyst

The $Fe_3O_4$_PDA_Lipase was recovered by magnetic decantation from both the biodiesel and glycerol layers. Separated $Fe_3O_4$_PDA_Lipase was then washed and dried at ambient temperature for reuse.

### 2.9. Characterization of Biodiesel

Biodiesel produced by $Fe_3O_4$_PDA_Lipase catalyzed transesterification was monitored by Fourier transfer infrared (FTIR) spectroscopy. The FTIR spectra of feedstock and synthesized biodiesel were taken over the scanning range 400–4000 $cm^{-1}$ using Cary 630 Agilent FTIR spectrometer. Oil FTIR spectra was compared to the FTIR of synthesized biodiesel to confirm the formation of fatty acid methyl esters. The GC/MS scan of biodiesel was taken to reveal compositional profile of synthesized biodiesel. GCMS QP 2010 system with a dB 5 column and a diameter of 0.15 mm was used. One μL sample size and 1:100 split ratio was selected. Helium as the carrier gas at 1.2 mL/min rate was used to elute the sample. The column temperature was set in the range 150–250 °C, at rate of 4 °C/min. GCMS mass scanning range was 30 to 550 m/z. Detection of FAMEs was done using NIST MS library of GCMS. The compatibility of WCO based biodiesel being used as fuel was confirmed by the estimation of its fuel properties, which were investigated according to the ASTM D methods.

### *2.10. Selection of Suitable Models for Optimization*

Based on the experimental outputs, the best fitted model out of linear, 2Fi, cubical and quadratic models were selected for optimization purpose. The fitness of selected models was further ascertained through summery statistics such as $R^2$, adjusted $R^2$, model significance and lack-of-fit test. Moreover, normality and predicted values versus actual values plots were also employed for the above said purpose. The impact of selected reaction variables on the response i.e., biodiesel yield was revealed by 3D response surface plots.

## 3. Results and Discussion

### *3.1. Synthesis and Evaluation of Nano-Biocatalyst Effectiveness*

#### 3.1.1. XRD Analysis

The XRD pattern of $Fe_3O_4$ nanoparticles formed by solvothermal method is shown in Figure 2. The XRD pattern of formed nanoparticles resembles with JCPDS card No# 019–0629 for magnetite, with no extraneous peak, which indicates the high purity of the magnetite. The 2-theta value of peaks were 29.95°, 35.05°, 42.73°, 53.37°, 56.63°, and 62.29°, which attribute to the interplanar spacing or d value in angstrom with their (hkl) of 2.98 (220), 2.56 (311), 2.12 (400), 1.72 (422), 1.62 (511) and 1.49 (440), respectively. The peak values resemble with characteristic XRD peaks of $Fe_3O_4$ lattice with cubic spinal shape. The sharp peaks correspond to the high crystallinity of the magnetite. The particle size calculated using Scherrer equation was 24.18 nm, which is close to the value calculated by TEM images showing homogeneity in the particle sizes. The result has resembled with previously reported results [29,30].

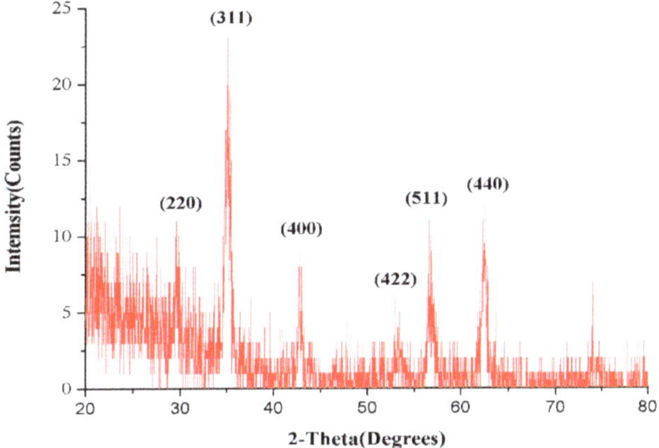


**Figure 2.** XRD (X-ray diffraction) graph of $Fe_3O_4$ nanoparticles.

### 3.1.2. TEM Analysis

TEM analysis was done for a detailed investigation of morphological characters. TEM images of the magnetic nanoparticles (Figure 3) were used to confirm the size and the shape of the nanoparticles. One can see from Figure 3a,b, that all the nanoparticles are homogeneous in size and shape. The size calculated using the TEM images ranges from 24 to 27 nm, whereas the shape is almost round. No agglomeration appeared in the micrograph, which may attribute to the homogeneous depression of these nanoparticles. TEM image (Figure 3c) shows that the nanoparticles have a mesoporous structure, which helps to achieve better catalytic properties.

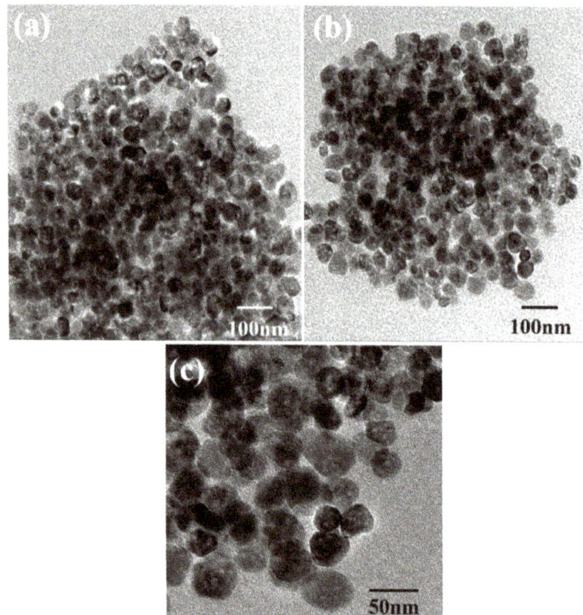


**Figure 3.** TEM images of the $Fe_3O_4$ nanoparticles at (**a**) 100 nm; (**b**) 100 nm and (**c**) 50 nm magnifications.

### 3.1.3. FTIR Analysis of Nanoparticles

A small amount of the product formed by the addition of dopamine in $Fe_3O_4$ suspension was taken out, washed with distilled water and dried using desiccator. The dried product was then subjected to Fourier transfer infrared spectroscopic analysis. FTIR confirms the attachment of dopamine on iron oxide nanoparticles by the presence of four peaks. Absorption band at 3209 $cm^{-1}$ corresponds to the characteristic starching frequency of hydroxyl groups merged with a starching frequency band of N-H. This broad band was due to catechol groups of dopamine at the surface. Bending vibration peak of N-H was at 1628 $cm^{-1}$. Sharp peak at 1290 $cm^{-1}$ indicated the presence of C-O bond. While peak at 1472 $cm^{-1}$ was attributed to the C=C ring starching band overlapped with $-CH_2$ scissoring band (Figure 4). The results were depicted to be comparable to previous research [24].

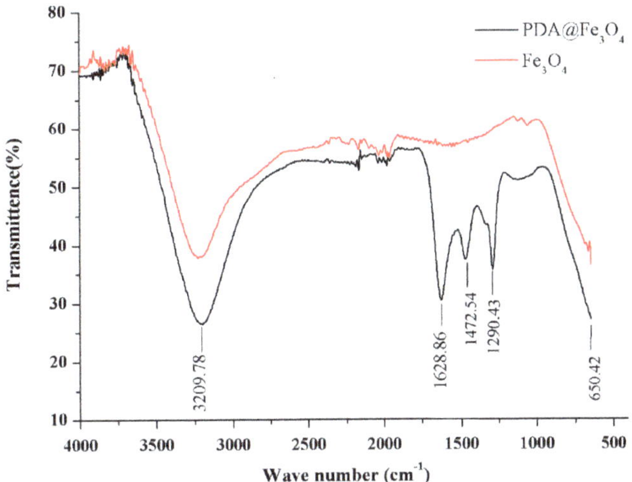


**Figure 4.** Comparative FTIR of naked $Fe_3O_4$ and polydopamine coated $Fe_3O_4$ nanoparticles.

### 3.1.4. Lipase Activity Assay

The activity titer of free lipase was found to be 18.32 U/mg/min, while the activity titer of $Fe_3O_4$_PDA_Lipase was found to be 17.83 U/mg/min. Lipase loading of 0.336 g (84.2% of used enzyme) was achieved by using polydopamine functionalized $Fe_3O_4$. The results showed that polydopamine gave high lipase loading and lipase activity because the complex formed by the combination of polydopamine and $Fe_3O_4$ nanoparticles was found to be efficient regarding the immobilization of lipase providing wider surface during the reactions, hence gave higher conversion rate in a short period of time [31].

### 3.1.5. Effect of pH on Nano-Biocatalyst Activity

The impact of pH on the activities of free and immobilized lipase was studied to find the optimum pH of enzyme activity. The effect of pH in the range of 5 to 10 was studied (Figure 5), and it was found that maximum lipase activity was exhibited at pH 7.0 for free lipase (i.e., 18.32 U/mg/min taken as 100% for free lipase) while lipase immobilized on $Fe_3O_4$ nanoparticles gave maximum activity at pH 8 (i.e., 17.83 U/mg/min taken as 100% for immobilized lipase). Above and below this pH, lipase activity was reduced. The results showed that, the relationship curve between pH and activity of lipase immobilized on $Fe_3O_4$ nanoparticles was shifted towards the right, which demonstrate that due to lipase immobilization on functionalized magnetic nanoparticles, the ability of lipase tolerance increased even if the pH of the reaction mixture varied. This means that by the immobilization of lipase, its adaptability to a wide pH range increased as compared to free lipase. Our studies were in accordance with the studies of Baharfar and Mahajer, in which magnetic nanoparticles were used for the immobilization of lipase, and immobilized lipase gave maximum activity at increased pH as compared to free lipase [31].

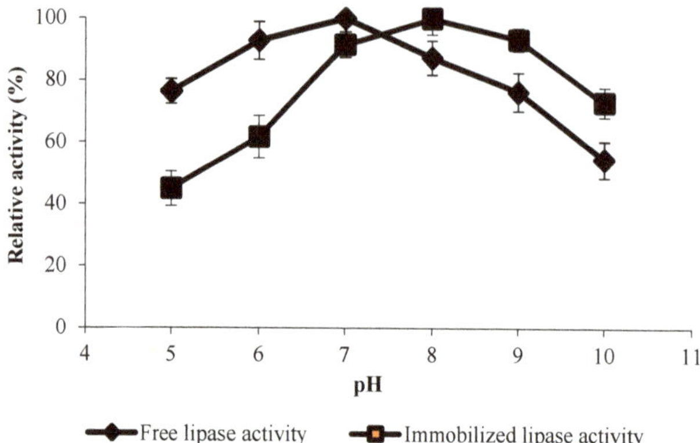


**Figure 5.** Effect of pH on the activities of free and immobilized lipase (date points are the mean of triplicate reaction values).

#### 3.1.6. Effect of Temperature on Activity of Nano-Biocatalyst

The impact of temperature on the activity of free and immobilized lipase was studied in the range of 25 to 50 °C (Figure 6) and it was depicted that maximum activity of free lipase was obtained at 37 °C, while the lipase immobilized on $Fe_3O_4$ nanoparticles showed maximum activity at the temperature of 40 °C, which means that lipase immobilized on nanoparticles was tolerant to high temperature fluctuations and was stable at a wider temperature range. The covalent bonds formed during immobilization may have increased the stability and tolerability of the lipase. Similar studies were reported by other researchers [23,31].

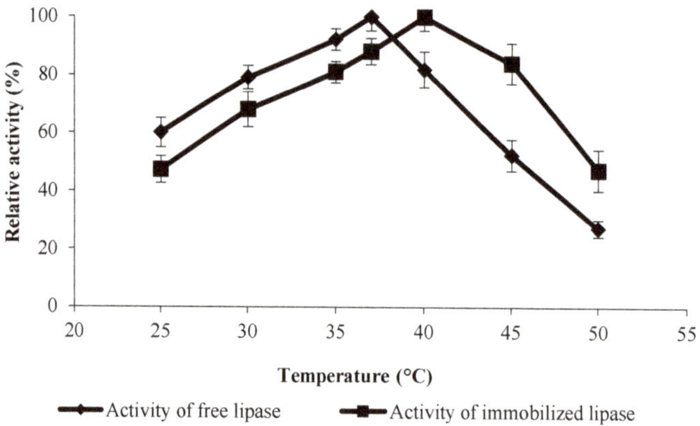


**Figure 6.** Effect of temperature on the activities of free and immobilized lipase (date points are the mean of triplicate reaction values).

### 3.2. Process Optimization/Characterization

#### 3.2.1. Characteristics of Feedstock

The density and the specific gravity of feedstock were 0.918 ± 0.050 g/cm$^3$ and 0.934 ± 0.051 respectively, while the acid value was calculated to be 1.68 ± 0.12 mg/KOHg, the peroxide value was 15.80 ± 0.32 mEq/Kg, iodine value was 96.20 ± 1.00 and the saponification value of waste cooking oil feedstock was 182.0 ± 1.0 mg KOH/g.

### 3.2.2. Composition Analysis of Produced Biodiesel by GC-MS

The product (biodiesel) was analysed by GC-MS. Major fatty acid methyl esters consisted of methyl esters of palmitic acid (16:0), lenoleic acid (18:2), stearic acid (18:0), gonodic acid (20:1) and arachidic acid (20:0) were identified by NIST library of GC-MS. Noshadi et al. [32] reported waste cooking oil with myristic acid, palmitoleic acid, palmitic acid, stearic acid, oleic acid, linoleic acid and linolenic acid. Uzun et al. [33] used waste oils for biodiesel production with a following fatty acid content of palmitic acid, stearic acid, oleic acid, linoleic acid and docasanoic acid. Comparable results have also been reported by [34–36], who found that the variations in fatty acid content might be due to presence of different edible oils and their varying amounts in cooking oils.

### 3.2.3. FTIR Monitoring of Biodiesel Production

Conversion of triglycerides into the biodiesel was confirmed by the FTIR spectral comparison of feedstock oil with the synthesized biodiesel. Characteristic peaks of waste cooking oil were observed at 723, 1700–1800 and 2800–3000 $cm^{-1}$, which corresponds to $-CH_2$ rocking, C=O starching and symmetric C-H stretching vibrations, respectively. While in the case of biodiesel, adsorption peak at 1437 $cm^{-1}$ corresponding to methyl ester and peak appearing at 1196 $cm^{-1}$ for C-O ester bond are the characteristic biodiesel adsorption peaks. The absence of 1380 $cm^{-1}$ peak in biodiesel that represents the O–$CH_2$ bonds in glycerol site of triglycerides further proves the formation of methyl esters (Figure 7). The results obtained in the present work are comparable to previous research [37–39].

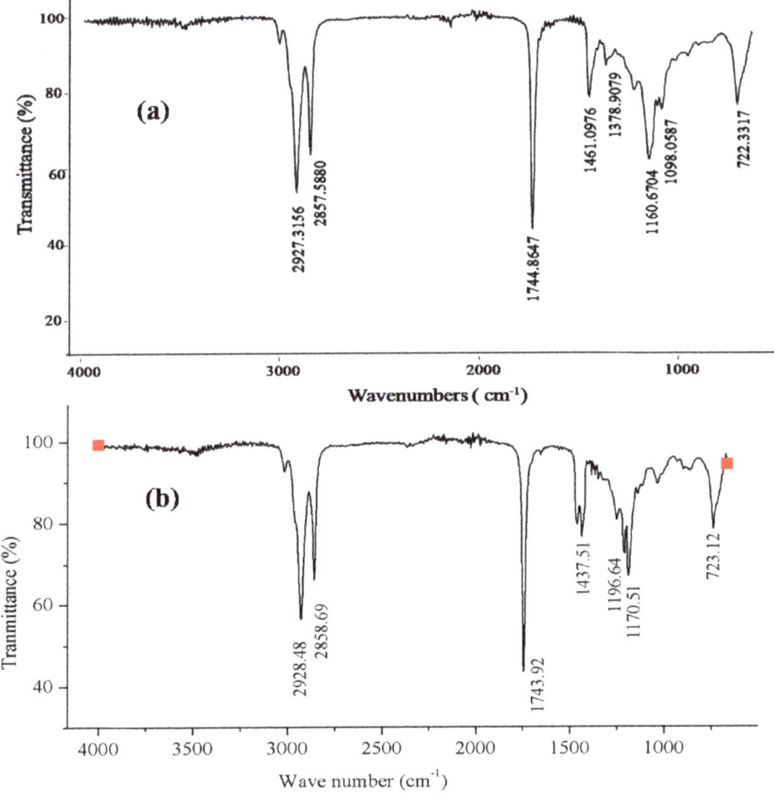


**Figure 7.** FTIR of (**a**) WCO and (**b**) WCO based biodiesel.

### 3.2.4. Optimized Reaction Parameters

Table 1 describes the optimum reaction conditions for the optimized yield of waste cooking oil-based biodiesel. By carrying out 50 experiments on various conditions of selected parameters, it has been observed that $Fe_3O_4$_PDA_Lipase catalysed the transesterification of feedstock oil giving a maximum yield of 92%, under the given set of reaction conditions/parameters viz., biocatalyst concentration (10%), $CH_3OH$ to oil molar ratio (6:1), reaction time (30 h), reaction temperature (37 °C) and water content (0.6%). While biocatalyst concentration (1%), methanol to oil molar ratio (3:1), reaction time (12 h), reaction temperature (20 °C) and water content (1%) gave the lowest response (41.5%).

Stoichiometrically, methanol reacts with triglycerides in 3:1 molar ratio, but excess amount of methanol was provided to increase the rate of forward reaction because transesterification is equilibrium limited reaction [40]. Moreover, the viscosity of the oils is high, which hinders the mass transfer. The addition of excess methanol lessens the reaction mixture viscosity, so the rate of reaction is increased by the improved mass transfer [41]. However, excess of methanol can lead to the emulsification of glycerol, which can recombine with the esters to reduce the biodiesel yield. In addition, excess amount of methanol can deactivate the lipase by changing its globular structure [42,43]. In the current study, a 6:1 methanol to oil ratio was observed to be the optimum $CH_3OH$ concentration for the biocatalytic transesterification of waste cooking oil.

Water presence influences the activity of the immobilized enzyme and stability of the free enzyme. Water/oil interface may also be required for better catalytic activity of lipase. Furthermore, the enzyme deactivation due to small chain alcohols can be prevented if the lipase is dispersed in water [43]. However, in the presence of higher content of water hydrolysis may compete with the methanolysis [44]. Therefore, water content is required to be adjusted with minimal content as per experimental requirements. Highest triglyceride conversion into biodiesel was obtained while using 0.6% $H_2O$ content in present research. Temperature significantly effects the enzymatic transesterification. The activity of lipase increases with temperature till the optimum point, i.e., 37 °C, and a further increase in temperature can lead to denaturisation of lipase.

Similar conditions have also been investigated by other researchers. Ying and Dong obtained the maximum biodiesel yield using immobilized lipase with 31.3% nano-biocatalyst, 38 °C temperature and 4.7:1 methanol to oil ratio [27]. Thangaraj et al. reported the maximum biodiesel yield of 89% at an optimum condition i.e., 1:3 methanol/oil ratio, 12 h of reaction period and 45 °C temp., using the immobilized NS81006 lipase as the catalyst [45]. Iso et al. reported the highest enzyme activity at 0.3% water content using immobilized lipase [46]. Mumtaz et al., using enzymatic transesterification, reported 95.9% biodiesel yield with 0.75% NOVOZYME-435, 6:1 methanol to oil ratio and 60 h of reaction at 32.50 °C [47]. Arumugam and Ponnusami used lipase immobilized on activated carbon for biodiesel production; optimum parameters were 1:9 oil to methanol ratio, 10% water content and 30 °C temperature, achieving a yield of 94.5% biodiesel with these conditions [41]. Xia also used RSM to optimize the enzymatic transesterification process, the optimum conditions reported are 4:1 methanol to oil ratio, 6.8% of biocatalyst and 42.2 °C reaction temperature [26]. Andrade et al. used nano-biocatalyst prepared by immobilization of lipase on magnetic nano-particles and achieved the highest biodiesel yield at 4:1 methanol/oil raio, 20% enzyme conc., 37 °C temp., and 12 h of time period [25]. Andrade et al. attained 95.5% methyl ester yield at 8:1 molar ratio of alcohol to oil, 8% biocatalyst amount and 45 °C temperature [48].

**Table 1.** Optimum reaction conditions for biodiesel production from $Fe_3O_4$-PDA-Liapase catalysed transesterification.

| Feedstock | Catalyst | Catalyst Conc. (%) | Methanol to Oil Ratio (%) | Reaction Time (h) | Reaction Temp. (°C) | Water Conc. (%) | Biodiesel Yield (%) |
|---|---|---|---|---|---|---|---|
| WCO | $Fe_3O_4$_PDA_Lipase | 10 | 6:1 | 30 | 37 | 0.6 | 92 |

### 3.2.5. Model Validation

Triplicate reactions were executed at optimum conditions to validate model accuracy. The average yield of 92.0 ± 1.3% was revealed to be comparable with that of predicted value, which was 88.93%. Hence, the model was depicted to be accurate and valid to estimate the response, i.e., yield of biodiesel.

### 3.2.6. ANOVA for Response Surface Quadratic Models for Biodiesel Production

For biodiesel production catalysed by $Fe_3O_4$_PDA_Lipase, quadratic model was depicted to be most suitable one with $p$-value < 0.05. The adjusted $R^2$ value for quadratic model was found to be 0.9840, while the lack of fit value was 0.0531. Insignificant lack of fit test value in addition to $R^2$ values also suggested/ascertained the fitness of quadratic model for $Fe_3O_4$_PDA_Lipase catalysed transesterification reactions.

Table 2 presents ANOVA describing selected model significance along with the significance of understudy reaction conditions. Out of the linear terms; A—methanol to oil ratio, B—enzyme concentration, C—reaction temperature and D—reaction time were ascertained to be significant, having $p$-values less than 0.05. While $p$-value i.e., 0.0874 > 0.05 depicted non-significant impact of E—water concentration. The above said factors significantly affected to the response as linear interactions i.e., AB, AC, AD, BD, and BE with $p$-values of 0.0010, 0.0321, 0.0076, 0.0008 and 0.0201 < 0.05, respectively. $A^2$, $B^2$ and $C^2$ were found to have significant effect with $p$-values 0.0245, < 0.0001 and < 0.0001, whilst $D^2$ and $E^2$ were insignificant quadratic term.

Model equation in terms of coded values is as follow;

$$\text{Biodiesel yield} = +86.03 + 1.97\,A + 14.73\,B + 0.97\,C + 2.24\,D - 0.68\,E + 1.45\,AB - 0.89\,AC - 1.13\,AD + 0.76\,AE + 0.51\,BC + 1.47\,BD - 0.97\,BE - 0.57\,CD + 0.22\,CE + 0.053\,DE - 3.37\,A^2 - 11.87\,B^2 - 8.90\,C^2 - 1.57\,D^2 - 1.72\,E^2$$

By the comparison of the factor's coefficients in the equation, the relative impact of each factor on the biodiesel yield can be identified.

**Table 2.** ANOVA for response surface quadratic model for biodiesel yield.

| Source | Df | Sum of Squares | Mean Square | F Value | $p$-Value Prob > F |
|---|---|---|---|---|---|
| Model | 20 | 15,145.61 | 757.28 | 151.53 | <0.0001 |
| A—methanol to oil ratio | 1 | 132.56 | 132.56 | 26.53 | <0.0001 |
| B—enzyme concentration | 1 | 7374.92 | 7374.92 | 1475.73 | <0.0001 |
| C—reaction temperature | 1 | 32.22 | 32.22 | 6.45 | 0.0167 |
| D—reaction time | 1 | 170.11 | 170.11 | 34.04 | <0.0001 |
| E—water | 1 | 15.64 | 15.64 | 3.13 | 0.0874 |
| AB | 1 | 67.57 | 67.57 | 13.52 | 0.0010 |
| AC | 1 | 25.34 | 25.34 | 5.07 | 0.0321 |
| AD | 1 | 41.18 | 41.18 | 8.24 | 0.0076 |
| AE | 1 | 18.45 | 18.45 | 3.69 | 0.0645 |
| BC | 1 | 8.44 | 8.44 | 1.69 | 0.2041 |
| BD | 1 | 69.33 | 69.33 | 13.87 | 0.0008 |
| BE | 1 | 30.23 | 30.23 | 6.05 | 0.0201 |
| CD | 1 | 10.59 | 10.59 | 2.12 | 0.1563 |
| CE | 1 | 1.61 | 1.61 | 0.32 | 0.5744 |
| DE | 1 | 0.090 | 0.090 | 0.018 | 0.8940 |
| $A^2$ | 1 | 28.12 | 28.12 | 5.63 | 0.0245 |
| $B^2$ | 1 | 348.60 | 348.60 | 69.76 | <0.0001 |
| $C^2$ | 1 | 188.78 | 188.78 | 37.78 | <0.0001 |
| $D^2$ | 1 | 6.11 | 6.11 | 1.22 | 0.2779 |
| $E^2$ | 1 | 7.33 | 7.33 | 1.47 | 0.2355 |
| Residual | 29 | 144.93 | 5.00 | | |
| Lack of Fit | 22 | 132.35 | 6.02 | 3.35 | 0.0531 |
| Pure Error | 7 | 12.58 | 1.80 | | |
| Cor Total | 49 | 15,290.53 | | | |

Li and Yan [49] applied three factor RSM for the optimization of transesterification process using immobilized lipase as catalyst, enzyme conc. and temp. were ascertained to be significant linear terms, but methanol to oil ratio was insignificant term with (*p*-value of 0.3083) all quadratic terms were also significant, while no first order interaction term was significant (Table 2). Wu et al. [50] optimized four reaction parameters using RSM, which were (i) lipase concentration, (ii) reaction time, (iii) reaction temperature, and (iv) ethanol to oil molar ratio. Lipase level, temperature and time were revealed to be the significant linear terms that support the present study, while ethanol to oil ratio showed insignificant impact, which might be due to the shorter range (3:1 to 6:1) of alcohol: oil used in that design. (Time × Temperature) was the only significant first order interaction terms. Huang et al. [51] reported a significant quadratic model for the optimization of methyl ester formation. Lipase to oil ratio ($x_1$), ratio of two lipases ($x_2$), t-butanol:oil ($x_3$), methanol:oil ($x_4$) and time ($x_5$) were independent variables chosen for optimization of reaction. Out of that, all the linear terms showed significant effect, while $x_1x_2$, $x_2x_3$, $x_3x_4$, $x_2x_4$, $x_1x_4$, and $x_4x_5$ were significant first order terms and among quadratic terms $x_3^2$ and $x_4^2$ had significant effect on response. Li and Dong [27] selected methanol:oil $x_1$, lipase:oil $x_2$, water content $x_3$, and temperature $x_4$ as independent variables for process optimization of biodiesel production using RSM. All the linear terms were significant in that model as well, which resembles the current study except the water content, which is insignificant in our work. All quadratic and first order interaction terms were also significant, expect for $x_1x_4$ and $x_2x_4$, and similar first order terms are insignificant in the present work as well. Xia [26] reported methanol:oil, amount of enzyme, time for reaction and amount of hexane (solvent) as significant linear terms, all the quadratic terms showed significant impact/effect on response, while (enzyme × hexane content) was the only significant first order interaction term. The results of ANOVA for the optimization of biodiesel production using immobilized lipase are comparable to the literature; the few variations might be due to different fatty acid profiles, different ranges of selected parameters, different alcohols being used and because of using solvent/solvent free systems.

Predicted vs. actual plot (Figure 8) of % biodiesel yield for the design, depict the fitness of the quadratic model, the difference between the predicted and the actual values are very small, as presented in Figure 8, which confirms the fitness of quadratic model. Our findings are comparable to the previous reports [27,51].

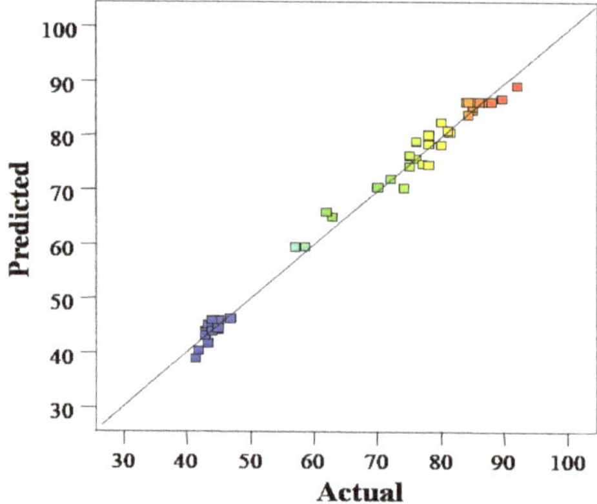


**Figure 8.** Predicted vs. actual values plot.

### 3.2.7. Response Surface (RS) Plots of Interacting Terms

RS plots of the significant first order interaction terms are presented in Figure 9. Figure 9a is the RS plot for biocatalyst ($Fe_3O_4$_PDA_Lipase) concentration and methanol to oil ratio. 3D plot revealed that the biodiesel yield increases with the increase in biocatalyst concentration and the content of methanol as the highest biodiesel yield is obtained at a biocatalyst concentration of 10% and 6:1 methanol to oil ratio, with further increase in the methanol content the yield starts to decrease, which might be due to deactivation of the enzyme by excess methanol. Figure 9b indicates the effect of methanol to oil ratio and reaction temperature interaction. The highest response value at the centre shows that the yield increases with temperature and methanol to oil ratio till the optimum points, then increase in temperature and methanol content results in decreased in biodiesel yield. Figure 9c 3D plot describes the influence of reaction time and methanol: oil ratio on response. Figure 9d is the response surface plot of enzyme concentration (%) and reaction time (h). The highest response at the inner corner presents that the biodiesel yield increases with both reaction time and the biocatalyst concentration, as the highest yield is obtained at 10% biocatalyst concentration and 30 h of reaction time. BE (enzyme conc. × water content) is another significant interaction term. The enzyme activity and structure are affected by the water content, which is clearly indicated by Figure 9e, which predicts that the biodiesel yield decreases by reduced enzyme activity as the water percentage deviates from optimum value.

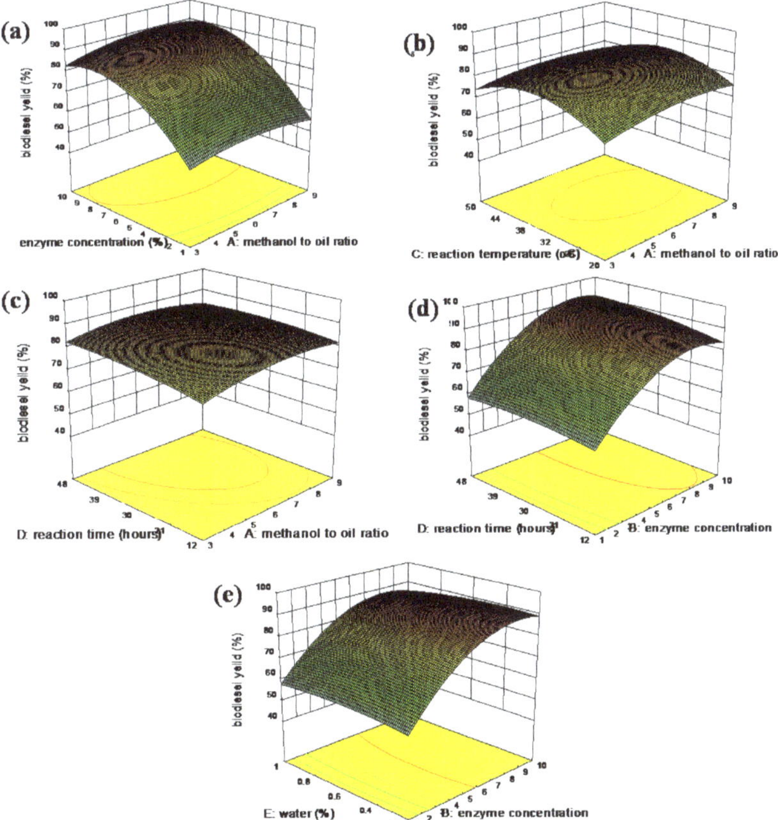


**Figure 9.** Response surface graphs of significant first order interaction terms (**a**) A × B, (**b**) A × C, (**c**) A × D, (**d**) B × D and (**e**) B × E.

### 3.2.8. Recovery and Reusability of Nano-Biocatalyst

At reaction completion, $Fe_3O_4$_PDA_Lipase was recovered by using magnetic decantation. Afterwards, lipase assay was carried out to find out the activity of immobilized lipase after recovery. No change in the activity of lipase was detected after first use, i.e., 17.83 U/mg/min. Therefore, these recovered $Fe_3O_4$ magnetic nanoparticles were reused several times for the biodiesel production and after the completion of each reaction the lipase activity assay was performed, which showed that lipase activity started decreasing after four uses, and after seven uses the activity declined to 4.6 U/mg/min. The biodiesel conversion rate decreased after four uses (Figure 10), which was clearly due to a decrease in the activity of immobilized lipase. This reduction in activity may be due to the exposure of nano-biocatalyst to organic solvents present in the reaction mixture or repeated exposure to heat, which may have resulted in the decrease of lipase activity. Similar studies have also been reported by Dumri and Hung [23].

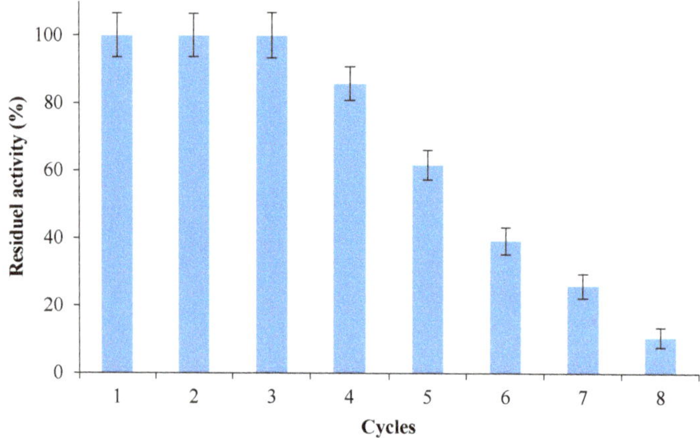


**Figure 10.** Re-usability of lipase immobilized $Fe_3O_4$ nanoparticles.

### 3.2.9. Physical Properties of Biodiesel

Fuel properties of produced biodiesel are presented in Table 3. The fuel properties meet the biodiesel standards set by ASTMD.

**Table 3.** Properties of WCO based biodiesel.

| Property | Unit | Value | ASTM D | Method |
|---|---|---|---|---|
| Acid value | (mg KOH/g) | 0.21 ± 0.06 | <0.50 | ASTM D 664 |
| Kinematic viscosity (40 °C) | ($mm^2$/s) | 4.10 ± 0.04 | 1.9–6.0 | ASTM D 445 |
| Flash point | (°C) | 170.41 ± 0.47 | >130 | ASTM D 93 |
| Pour point | (°C) | −3.0 ± 0.5 | - | ASTM D 97 |
| Density | (g $mL^{-1}$) | 0.871 ± 0.320 | 0.86–0.90 | ASTM D 5002 |

For the complete combustion of biodiesel, optimum fuel to air ratio is required, which is obtained at specific levels of density. Kinematic viscosity is a measure of restriction between the two layers of a liquid, and quality of combustion is recognized to be affected by kinematic viscosity. The ignition point of fuel on exposure to the spark is called the flash point. High flash points of the biodiesel make it easy to store. Pour point and cloud point are important for cold flow properties. Furthermore, suitability of biodiesel for the use in engine can also be evaluated by these properties. Previous studies have revealed comparable results for the biodiesel produced from WCO. Few variations in the observed fuel characteristics may be attributed to different source and composition of feedstock oil [33].

## 4. Conclusions

$Fe_3O_4$_PDA_Lipase nano-biocatalyst was successfully synthesized and utilized to produce biodiesel using waste cooking oil as feedstock. Using a best fitted quadratic model, different reaction conditions of biodiesel production process were optimized, and it was revealed that $Fe_3O_4$_PDA_Lipase actively catalysed the reaction, resulting in a highlight biodiesel yield i.e., 92%. The synthesized nano-biocatalyst proved to have better adoptability/stability and was easily recoverable using external magnet and retained most of its catalytic activity for four cycles. The properties of synthesized biodiesel were in agreement with the ASTM D standards. Conclusively, $Fe_3O_4$_PDA_Lipase has been proved to be a sustainable and cost-effective catalyst for biodiesel production.

**Author Contributions:** The conceptualization of this journal article is from T.T., M.W.M. and U.R.; the methodology was designed by T.T., M.W.M. and U.R.; the writing—original draft was prepared by T.T.; S.A., A.S. and M.W.M. helped with the editing and supervision. H.M. provided technical assistance regarding the synthesis and characterization of nano-biocatalyst, A.I., T.S.Y.C. and I.A.N. reviewed the manuscript for final version improvement. All authors have read and agreed to the published version of the manuscript.

**Funding:** The authors acknowledge their gratitude to King Saud University (Riyadh, Saudi Arabia) for the support of this research through Researchers Supporting Project number (RSP-2019/80).

**Conflicts of Interest:** The authors declare no conflict of interest.

## References


1. Shirneshan, A. HC, CO, $CO_2$ and $NO_x$ emission evaluation of a diesel engine fueled with waste frying oil methyl ester. *Procedia-Soc. Behav. Sci.* **2013**, *75*, 292–297. [CrossRef]
2. Srivastava, A.; Prasad, R. Triglycerides-based diesel fuels. *Renew. Sustain. Energy Rev.* **2000**, *4*, 111–133. [CrossRef]
3. Mehrasbi, M.R.; Mohammadi, J.; Peyda, M.; Mohammadi, M. Covalent immobilization of Candida antarctica lipase on core-shell magnetic nanoparticles for production of biodiesel from waste cooking oil. *Renew. Energy* **2017**, *101*, 593–602. [CrossRef]
4. Gui, M.M.; Lee, K.T.; Bhatia, S. Feasibility of edible oil vs. non-edible oil vs. waste edible oil as biodiesel feedstock. *Energy* **2008**, *33*, 1646–1653. [CrossRef]
5. Aladedunye, F.A.; Przybylski, R. Degradation and Nutritional Quality Changes of Oil During Frying. *J. Am. Oil Chem. Soc.* **2008**, *86*, 149–156. [CrossRef]
6. Paul, S.; Mittal, G.S. Regulating the use of degraded oil/fat in deep-fat/oil food frying. *Crit. Rev. Food Sci. Nutr.* **1997**, *37*, 635–662. [CrossRef]
7. Raqeeb, M.A.; Bhargavi, R. Biodiesel production from waste cooking oil. *J. Chem. Pharm. Res.* **2015**, *7*, 670–681.
8. Zhang, H.; Ding, J.; Zhao, Z. Microwave assisted esterification of acidified oil from waste cooking oil by CERP/PES catalytic membrane for biodiesel production. *Bioresour. Technol.* **2012**, *123*, 72–77. [CrossRef]
9. Balat, M.; Balat, H. Progress in biodiesel processing. *Appl. Energy* **2010**, *87*, 1815–1835. [CrossRef]
10. Karmakar, A.; Karmakar, S.; Mukherjee, S. Properties of various plants and animals' feedstocks for biodiesel production. *Bioresour. Technol.* **2010**, *101*, 7201–7210. [CrossRef]
11. Fukuda, H.; Kondo, A.; Noda, H. Biodiesel fuel production by transesterification of oils. *J. Biosci. Bioeng.* **2001**, *92*, 405–416. [CrossRef]
12. Köse, Ö.; Tüter, M.; Aksoy, H.A. Immobilized Candida antarctica lipase-catalyzed alcoholysis of cotton seed oil in a solvent-free medium. *Bioresour. Technol.* **2002**, *83*, 125–129. [CrossRef]
13. Bisen, P.S.; Sanodiya, B.S.; Thakur, G.S.; Baghel, R.K.; Prasad, G. Biodiesel production with special emphasis on lipase-catalyzed transesterification. *Biotechnol. Lett.* **2010**, *32*, 1019–1030. [CrossRef] [PubMed]
14. Sheldon, R.A.; Van Pelt, S. Enzyme immobilisation in biocatalysis: Why, what and how. *Chem. Soc. Rev.* **2013**, *42*, 6223–6235. [CrossRef] [PubMed]
15. Al-Zuhair, S.; Dowaidar, A.; Kamal, H. Dynamic modeling of biodiesel production from simulated waste cooking oil using immobilized lipase. *Biochem. Eng. J.* **2009**, *44*, 256–262. [CrossRef]
16. Rossi, L.M.; Costa, N.J.S.; Silva, F.P.; Gonçalves, R.V. Magnetic nanocatalysts: Supported metal nanoparticles for catalytic applications. *Nanotechnol. Rev.* **2013**, *2*, 5. [CrossRef]
17. Matsunaga, T.; Kamiya, S. Use of magnetic particles isolated from magnetotactic bacteria for enzyme immobilization. *Appl. Microbiol. Biotechnol.* **1987**, *26*, 328–332. [CrossRef]
18. Wang, C.; Han, H.; Jiang, W.; Ding, X.; Li, Q.; Wang, Y. Immobilization of thermostable lipase QLM on core-shell structured polydopamine-coated $Fe_3O_4$ nanoparticles. *Catalysts* **2017**, *7*, 49. [CrossRef]

19. Adlercreutz, P. Immobilisation and application of lipases in organic media. *Chem. Soc. Rev.* **2013**, *42*, 6406–6436. [CrossRef]
20. Lee, H.; Rho, J.; Messersmith, P.B. Facile Conjugation of Biomolecules onto Surfaces via Mussel Adhesive Protein Inspired Coatings. *Adv. Mater.* **2009**, *21*, 431–434. [CrossRef]
21. Ren, Y.; Rivera, J.G.; He, L.; Kulkarni, H.; Lee, D.K.; Messersmith, P.B. Facile, high efficiency immobilization of lipase enzyme on magnetic iron oxide nanoparticles via a biomimetic coating. *BMC Biotechnol.* **2011**, *11*, 63. [CrossRef] [PubMed]
22. Guo, S.; Li, D.; Zhang, L.; Li, J.; Wang, E. Monodisperse mesoporous superparamagnetic single-crystal magnetite nanoparticles for drug delivery. *Biomaterials* **2009**, *30*, 1881–1889. [CrossRef] [PubMed]
23. Dumri, K.; Hung Anh, D. Immobilization of Lipase on Silver Nanoparticles via Adhesive Polydopamine for Biodiesel Production. *Enzyme Res.* **2014**, *2014*, 389739. [CrossRef] [PubMed]
24. Shabbir, A.; Mukhtar, H. Optimization process for enhanced extracellular lipases production from a new isolate of Aspergillus terreus ah-F2. *Pak. J. Bot.* **2018**, *50*, 1571–1578.
25. Andrade, M.F.C.; Parussulo, A.L.A.; Netto, C.G.C.M.; Andrade, L.H.; Toma, H.E. Lipase immobilized on polydopamine-coated magnetite nanoparticles for biodiesel production from soybean oil. *Biofuel Res. J.* **2016**, *3*, 403–409. [CrossRef]
26. Li, Y.X. Optimization of biodiesel production from rice bran oil via immobilized lipase catalysis. *Afr. J. Biotechnol.* **2011**, *10*, 16314–16324.
27. Li, Y.X.; Dong, B.X. Optimization of Lipase-Catalyzed Transesterification of Cotton Seed Oil for Biodiesel Production Using Response Surface Methodology. *Braz. Arch. Biol. Technol.* **2016**, *59*. [CrossRef]
28. Soltani, S.; Rashid, U.; Yunus, R.; Taufiq-Yap, Y.H.; Al-Resayes, S.I. Post-functionalization of polymeric mesoporous C@Zn core-shell spheres used for methyl ester production. *Renew. Energy* **2016**, *99*, 1235–1243. [CrossRef]
29. Cao, X.; Zhang, B.; Zhao, F.; Feng, L. Synthesis and properties of MPEG-coated superparamagnetic magnetite nanoparticles. *J. Nanomater.* **2012**, *2012*, 607296. [CrossRef]
30. Chen, D.; Tang, Q.; Li, X.; Zhou, X.; Zang, J.; Xue, W.-Q.; Xiang, J.-Y.; Guo, C.-Q. Biocompatibility of magnetic $Fe_3O_4$ nanoparticles and their cytotoxic effect on MCF-7 cells. *Int. J. Nanomed.* **2012**, *7*, 4973. [CrossRef]
31. Baharfar, R.; Mohajer, S. Synthesis and Characterization of Immobilized Lipase on $Fe_3O_4$ Nanoparticles as Nano biocatalyst for the Synthesis of Benzothiazepine and Spirobenzothiazine Chroman Derivatives. *Catal. Lett.* **2016**, *146*, 1729–1742. [CrossRef]
32. Noshadi, I.; Amin, N.A.S.; Parnas, R.S. Continuous production of biodiesel from waste cooking oil in a reactive distillation column catalyzed by solid heteropolyacid: Optimization using response surface methodology (RSM). *Fuel* **2012**, *94*, 156–164. [CrossRef]
33. Uzun, B.B.; Kılıç, M.; Özbay, N.; Pütün, A.E.; Pütün, E. Biodiesel production from waste frying oils: Optimization of reaction parameters and determination of fuel properties. *Energy* **2012**, *44*, 347–351. [CrossRef]
34. Kannan, G.; Anand, R. Effect of injection pressure and injection timing on DI diesel engine fuelled with biodiesel from waste cooking oil. *Biomass Bioenergy* **2012**, *46*, 343–352. [CrossRef]
35. Omar, W.N.N.W.; Amin, N.A.S. Optimization of heterogeneous biodiesel production from waste cooking palm oil via response surface methodology. *Biomass Bioenergy* **2011**, *35*, 1329–1338. [CrossRef]
36. Patil, P.D.; Gude, V.G.; Reddy, H.K.; Muppaneni, T.; Deng, S. Biodiesel production from waste cooking oil using sulfuric acid and microwave irradiation processes. *J. Environ. Protect.* **2012**, *3*, 107. [CrossRef]
37. Dubé, M.A.; Zheng, S.; McLean, D.D.; Kates, M. A comparison of attenuated total reflectance-FTIR spectroscopy and GPC for monitoring biodiesel production. *J. Am. Oil Chem. Soc.* **2004**, *81*, 599–603. [CrossRef]
38. Siatis, N.; Kimbaris, A.; Pappas, C.; Tarantilis, P.; Polissiou, M. Improvement of biodiesel production based on the application of ultrasound: Monitoring of the procedure by FTIR spectroscopy. *J. Am. Oil Chem. Soc.* **2006**, *83*, 53–57. [CrossRef]
39. Soares, I.P.; Rezende, T.F.; Silva, R.C.; Castro, E.V.R.; Fortes, I.C. Multivariate calibration by variable selection for blends of raw soybean oil/biodiesel from different sources using Fourier transform infrared spectroscopy (FTIR) spectra data. *Energy Fuel* **2008**, *22*, 2079–2083. [CrossRef]

40. Stergiou, P.Y.; Foukis, A.; Filippou, M.; Koukouritaki, M.; Parapouli, M.; Theodorou, L.G.; Hatziloukas, E.; Afendra, A.; Pandey, A.; Papamichael, E.M. Advances in lipase-catalyzed esterification reactions. *Biotechnol. Adv.* **2013**, *31*, 1846–1859. [CrossRef]
41. Arumugam, A.; Ponnusami, V. Production of biodiesel by enzymatic transesterification of waste sardine oil and evaluation of its engine performance. *Heliyon* **2017**, 3, e00486. [CrossRef] [PubMed]
42. Christopher, L.P.; Kumar, H.; Zambare, V.P. Enzymatic biodiesel: Challenges and opportunities. *Appl. Energy* **2014**, *119*, 497–520. [CrossRef]
43. Kulschewski, T.; Sasso, F.; Secundo, F.; Lotti, M.; Pleiss, J. Molecular mechanism of deactivation of *C. antarctica* lipase B by methanol. *J. Biotechnol.* **2013**, *168*, 462–469. [CrossRef] [PubMed]
44. Zhao, X.; Qi, F.; Yuan, C.; Du, W.; Liu, D. Lipase-catalyzed process for biodiesel production: Enzyme immobilization, process simulation and optimization. *Renew. Sustain. Energy Rev.* **2015**, *44*, 182–197. [CrossRef]
45. Thangaraj, B.; Jia, Z.; Dai, L.; Liu, D.; Du, W. Lipase NS81006 immobilized on $Fe_3O_4$ magnetic nanoparticles for biodiesel production. *Ovidius Univ. Ann. Chem.* **2016**, *27*, 13–21. [CrossRef]
46. Iso, M.; Chen, B.; Eguchi, M.; Kudo, T.; Shrestha, S. Production of biodiesel fuel from triglycerides and alcohol using immobilized lipase. *J. Mol. Catal. B Enzymat.* **2001**, *16*, 53–58. [CrossRef]
47. Mumtaz, M.W.; Adnan, A.; Mahmood, Z.; Mukhtar, H.; Malik, M.F.; Qureshi, F.A.; Raza, A. Biodiesel from Waste Cooking Oil: Optimization of Production and Monitoring of Exhaust Emission Levels From its Combustion in a Diesel Engine. *Int. J. Green Energy* **2012**, *9*, 685–701. [CrossRef]
48. Andrade, I.C.; Vargas, F.E.S.; Fajardo, C.A.G. Biodiesel production from waste cooking oil by enzymatic catalysis process. *J. Chem. Eng.* **2013**, *7*, 993–1000.
49. Li, Q.; Yan, Y. Production of biodiesel catalyzed by immobilized Pseudomonas cepacia lipase from Sapiumsebiferum oil in micro-aqueous phase. *Appl. Energy* **2010**, *87*, 3148–3154. [CrossRef]
50. Wu, W.; Foglia, T.; Marmer, W.; Phillips, J. Optimizing production of ethyl esters of grease using 95% ethanol by response surface methodology. *J. Am. Oil Chem. Soc.* **1999**, *76*, 517–521. [CrossRef]
51. Huang, Y.; Zheng, H.; Yan, Y. Optimization of lipase-catalyzed transesterification of lard for biodiesel production using response surface methodology. *Appl. Biochem. Biotechnol.* **2010**, *160*, 504–515. [CrossRef] [PubMed]

Article

# Diethyl Ether as an Oxygenated Additive for Fossil Diesel/Vegetable Oil Blends: Evaluation of Performance and Emission Quality of Triple Blends on a Diesel Engine

**Laura Aguado-Deblas [1], Jesús Hidalgo-Carrillo [1], Felipa M. Bautista [1], Diego Luna [1], Carlos Luna [1], Juan Calero [1], Alejandro Posadillo [2], Antonio A. Romero [1] and Rafael Estevez [1,*]**

1 Departamento de Química Orgánica, Universidad de Córdoba, Campus de Rabanales, Ed. Marie Curie, 14014 Córdoba, Spain; aguadolaura8@gmail.com (L.A.-D.); Jesus.hidalgo@uco.es (J.H.-C.); qo1baruf@uco.es (F.M.B.); diego.luna@uco.es (D.L.); qo2luduc@uco.es (C.L.); p72camaj@gmail.com (J.C.); qo1rorea@uco.es (A.A.R.)

2 Seneca Green Catalyst S.L., Campus de Rabanales, 14014 Córdoba, Spain; seneca@uco.es

* Correspondence: q72estor@uco.es; Tel.: +34-957212065

Received: 14 February 2020; Accepted: 23 March 2020; Published: 25 March 2020

**Abstract:** The aim of this work is to analyze the effect of using diethyl ether (DEE) as an oxygenated additive of straight vegetable oils (SVOs) in triple blends with fossil diesel, to be used in current compression ignition (C.I.) engines, in order to implement the current process of replacing fossil fuels with others of a renewable nature. The use of DEE is considered taking into account the favorable properties for blending with SVO and fossil diesel, such as its very low kinematic viscosity, high oxygen content, low autoignition temperature, broad flammability limits (it works as a cold start aid for engines), and very low values of cloud and pour point. Therefore, DEE can be used as a solvent of vegetable oils to reduce the viscosity of the blends and to improve cold flow properties. Besides, DEE is considered renewable, since it can be easily obtained from bioethanol, which is produced from biomass through a dehydration process. The vegetable oils evaluated in the mixtures with DEE were castor oil, which is inedible, and sunflower oil, used as a standard reference for waste cooking oil. In order to meet European petrodiesel standard EN 590, a study of the more relevant rheological properties of biofuels obtained from the DEE/vegetable oil double blends has been performed. The incorporation of fossil diesel to these double blends gives rise to diesel/DEE/vegetable oil triple blends, which exhibited suitable rheological properties to be able to operate in conventional diesel engines. These blends have been tested in a conventional diesel engine, operating as an electricity generator. The efficiency, consumption and smoke emissions in the engine have been measured. The results reveal that a substitution of fossil diesel up to 40% by volume can be achieved, independently of the SVO employed. Moreover, a significant reduction in the emission levels of pollutants and better cold flow properties has been also obtained with all blends tested.

**Keywords:** diethyl ether; castor oil; sunflower oil; straight vegetable oils (SVO); biofuel; diesel engine; electricity generator; smoke opacity; Bosch smoke number

## 1. Introduction

Nowadays, so many countries have established a climate and energy policy framework, advancing in decarbonization to arrive towards a new friendly climate economy [1]. In this respect, an unprecedented effort is being made to implement energy alternatives (photovoltaic, wind, hydrogen, and nuclear energy) that allow the gradual replacement of natural gas, coal and fossil fuels in the field of electricity generation for reducing the high consumption of fossil fuels [2,3]. However, there is no such equivalent in transport, since vehicles using fuel cells or electric motors cannot compete yet with fossil fuel engines, especially in the field of heavy trucks [4], aviation [5], or the shipping sector. In this context, the incorporation of biofuels as fossil fuel substitutes is the strategy assumed to accomplish this necessary energetic progress since there is no need of modifying the compression-ignition (CI) diesel engines of the current car fleet [6,7]. Therefore, the use of biofuels leads to easy and gradual integration into the worldwide transportation logistics systems.

Among the existing biofuels, biodiesel has emerged as one of the best options. Biodiesel is obtained through homogeneous or heterogeneous alkaline transesterification of vegetable oil or animal fat with methanol, giving rise to a mixture of mono-alkyl esters of long-chain fatty acids (FAMEs) [8,9]. Over the past decade, biodiesel has been of interest by contributing to reduce energy dependence on fossil fuels and to minimize greenhouse-gas emissions from transportation [10].

Despite the fact the replacement of fossil diesel by biodiesel seems easy, this process is still considered economically unfeasible. The main reason is the high production cost associated, among other things, to the purification process needed because of the generation of glycerol as a by-product (10wt% of the total biodiesel produced). Considering the economic difficulties in the production of biodiesel, the search for different alternative biofuels is still mandatory.

In this regard, the transformation of vegetable oils into high-quality diesel fuels, avoiding the glycerol formation, has been deeply investigated [11]. These biodiesel-like biofuels, such as Gliperol [12], DMC-BioD [13,14] or Ecodiesel [15,16], avoid the generation of residues or by-products, integrating glycerol in the reaction products. Thus, these new biofuels can be obtained as soluble derivatives in transesterification processes, analogous to that for the production of FAME [17]. Besides, it has been reported as a high-quality diesel fuel known as "green diesel´´, which can be obtained by several treatments, such as cracking, pyrolysis, hydrodeoxygenation, and hydrotreating of vegetable oils [18].

Very recently, the use of straight vegetable oils (SVO) has become an interesting option for the replacement of fossil diesel. On one hand, vegetable oils can be obtained easily from agricultural or industrial sources, avoiding energetic costs associated with the transesterification required to obtain biodiesel. On the other hand, all the relevant physicochemical properties of vegetable oils are analogous to conventional diesel, except for the viscosity, which is much higher in oils. The high viscosities cause poor fuel atomization by premature injector contamination. To solve this issue without the need for carrying out the transesterification reaction, researchers have focused their attention on the reduction of high-viscosity oils by blending them with low viscous biofuels. Thus, it is possible to obtain double blends that comply with the requirements stipulated in the current diesel engines (EN 590 standard). Several blends have been reported employing plant-based sources such as vegetable oils and organic compounds [19,20]. These compounds exhibit a relatively short carbon chain, low viscosity values and also low cetane number since they have been identified as LVLC (Low Viscosity Low Cetane) fuels [21].

Employing this strategy, low viscous vegetable oils, such as pine oil or camphor oil, have also been studied in several blends with different biodiesels [22,23], vegetable oils [19] or with fossil diesel [24–26], to improve performance in the CI engines. So far, the additives studied as LVLC are natural compounds (pine oil, eucalyptus oil, camphor oil or orange oil), obtained from crops, which may contribute to reducing the world's dependence on oil imports, providing benefits to local agricultural industries. Similarly, compounds obtained by chemical synthesis from renewable products, mainly alcohols (methanol, ethanol, and butanol), are also applied as LVLC [27–29]. This strategy has even been applied with a non-renewable compound such as gasoline, capable of reducing the viscosity of vegetable oils in double and triple blends with fossil diesel [30,31]. Following this methodology, high levels of fossil fuel substitution have been obtained in a technically and economically feasible way.

Therefore, if a renewable compound like diethyl ether (DEE) is employed, instead of a non-renewable one such as gasoline, we can make the process greener. In fact, DEE can be easily obtained from ethanol, which can be obtained from biomass. Despite the fact that DEE is known as a cold-start aid for engines, its potential as a transportation fuel in blends with vegetable oil and/or with fossil diesel has not been much investigated. This molecule has several favorable properties for blending with diesel fuel, including very low kinematic viscosity, low autoignition temperature, high oxygen content, broad flammability limits, high miscibility with vegetable oils and diesel fossil, and very low values of cloud point (CP) and pour point (PP) that improves cold flow properties [29]. Therefore, it is expected that by blending the DEE with SVO, a notable improvement of some of the fuel properties will be obtained, such as a reduction of the CP and PP and so on. However, the calorific power of this compound is relatively low, so this fact could limit the percentage of substitution of fossil fuel by the DEE/oil blend to operate in today's internal combustion diesel engines, maintaining the appropriate parameters of EN 590 standard. In fact, we have recently reported that the low calorific power of ethanol and 2-propanol (27 and 33 MJ/kg, respectively) constitutes the greatest limitation for its use in double blends with oils [32]. This calorific power is very similar to that for DEE (34 MJ/kg) and, therefore, it is foreseeable it will show similar behavior.

Another aspect to take into consideration is the very low ignition quality of some diesel/DEE blends [27] that promote a high ignition delay, given the low heat of evaporation value that DEE exhibits [33]. This ignition delay can be overcome by the addition of vegetable oils to the blend. Therefore, a mutual benefit can be obtained by the use of DEE with vegetable oils in triple blends with diesel, i.e., DEE reduces the high viscosity of the oils, whereas the oils could compensate for the heat and evaporation of DEE. In fact, DEE has been reported as a low-emission renewable fuel and high-quality combustion improver in blends with diesel fossil [27,28], with biodiesel [27,29,34,35], with oils [27,29] or with diesel/oil [36–38]. Besides, better performance of compression-ignition engines operating with DEE/diesel/biodiesel triple blends has been achieved [39,40].

Waste cooking oil (sunflower oil) and castor oil have been selected as the vegetable oils for this work since they come from crops that are not destined to human or livestock food and they are easily available, not competing with oils for food uses. Sunflower oil has been selected as a standard reference to study the behavior of waste cooking oils because the use of any waste cooking oil from different sources, implies the difficulty of reproducing the results obtained. Castor oil is currently the only inedible vegetable oil available on an industrial scale.

The present study intends to advance the strategy of the substitution of fossil fuels by others of a renewable nature that can be used in current diesel engines in a viable way, not only from a technical point of view but also economically and even more importantly, applicable in the most immediate way. To do so, DEE has been employed as an oxygenated additive in blends with diesel and vegetable oil. In this respect, the optimum proportion of DEE/vegetable oil blend will be evaluated based on the most significant parameter, the kinematic viscosity, to meet with appropriate parameters of EN 590 standard that allow its use in internal combustion diesel engines. Additionally, flow cold properties (cloud and pour points) will be studied to ascertain the applicability of the fuel in cold climates. The diesel/DEE/oil triple blends obtained have been tested in a diesel engine. The most important parameters, such as fuel consumption and power generation have been studied to know the viability of the new fuel produced. The degree of pollution of all blends will be also evaluated from the generated smoke opacity values.

## 2. Materials and Methods

Some of the significant physicochemical characteristics of diesel, diethyl ether, and SVOs (sunflower oil and castor oil) have been collected in Table 1.

**Table 1.** Properties of diesel, sunflower oil, castor oil, and diethyl ether. All data collected in the Table were taken from the literature [36,41–43], except for the kinematic viscosity values which were experimentally measured in this work.

| Property | Diesel | Sunflower Oil | Castor Oil | DEE |
|---|---|---|---|---|
| Density at 15 °C (kg/m$^3$) | 830 | 920 | 962 | 713 |
| Kinematic viscosity at 40 °C (cSt)[1] | 3.20 ± 0.01 | 37.80 ± 0.05 | 226.20 ± 0.05 | 0.21 ± 0.01 |
| Calorific value (MJ/kg) | 42.8 | 39.5 | 37.2 | 33.9 |
| Flash point (°C) | 66 | 220 | 228 | 16 |
| Auto-ignition temperature (°C) | 250 | 316 | 448 | 160 |
| Cetane number | 50 | 37 | 40 | >124 |

[1] Viscosity value errors were obtained from the average of 3 measurements.

### 2.1. Diethyl Ether/Vegetable Oil Double Blends, and Diesel/ Diethyl Ether /Vegetable Oil Triple Blends

Sunflower oil was bought from a local market and castor oil (Panreac, Castellar Del Valles, Spain) was purchased from a local commercial representative. DEE (≥99.5% purity) was procured from Sigma-Aldrich Chemical Company. First of all, sunflower oil (food grade) and castor oil were mixed with DEE in different concentrations to find out the optimum DEE/SVO double blends. The best double blends, which comply with the established requirements by European petrodiesel standard EN 590 for being employed as biofuels, were selected to be mixed with conventional diesel fuel (Repsol service station) in different proportions, from 20% to 100% by volume, denoted as B20, B40, B60, B80, and B100. The percentage of biofuel (DEE/SVO blend) added to fossil diesel is expressed as B, where B0 corresponding to 100% of fossil diesel and B100 means 100% of renewable DEE/oil biofuel. The components of all blends were manually mixed at room temperature. Additionally, all components were completely miscible with petroleum diesel, allowing the blending of these in any proportion. The obtained diesel/DEE/SVO triple blends were investigated as biofuels in this work.

### 2.2. Characterization of the Biofuel Blends

Fuel reformulation can affect the physicochemical and safety properties of the fuel. Hence, knowledge of these properties is especially important. In this work, some of the most crucial properties to evaluate the suitability of biofuels have been determined either experimentally or using specific equations to predict them.

As aforementioned, kinematic viscosity and cold flow properties are the rheological properties more influenced by blends of vegetable oils with fossil diesel and other additives. In fact, these properties play a crucial role in the correct performance of conventional diesel engines [32]. Kinematic viscosity significantly affects the quality of fuel atomization and the combustion process. Low viscosity can cause leakage in the fuel system while high viscosity can lead to incomplete combustion because of premature injector contamination [44].

Cold flow properties, such as cloud point and pour point, are responsible for solidification of fuel, causing operability problems as solidified material clogs fuel lines and filters. The temperature at which the crystals become visible (diameter ≥ 0.5 mm) is defined as the cloud point (CP), whereas the pour point (PP) is defined as the temperature at which the liquid ceases to flow. The CP usually occurs at higher temperatures than the PP [45]. Crystallization of fuel takes place when fuel molecules condensate forming a gel at low temperatures. Crystallization occurs in two general steps: The first step is the nucleation and crystal growth and the second step is the organization of the molecules creating a stable nucleus in a crystalline network. The continuous growth of the crystalline network generates the interruption of fuel flow, causing fuel starvation and incomplete combustion, which leads to starting problems in the vehicle in cold weather [46].

The flash point (FP) is a crucial property for production, handling, transportation and storage of fuels. This parameter provides an indication of the fire hazard of fuel under ambient conditions. The flash point is defined as the lowest temperature at which a liquid produces enough vapors to ignite in the presence of a flame or spark. Generally, a lower flash point is related to higher vapor pressure, so this property provides information on both flammability and volatility. The values of the flash point can be predicted from Kay's mixing rule:

$$T_{FP} = \sum_i yiTi \quad (1)$$

where T is the temperature corresponding to the flash point of the blend (°C), yi is the volume fraction of each component in the blend and Ti is the flash point of each component [47].

Another important property to define the efficiency of fuels is the calorific value (CV), also called the heat of combustion or calorific power, which is the quantity of heating energy released during complete combustion of a unit mass of the fuel, usually expressed in kilojoules per kilogram. The calorific value increases with increasing chain length and decreases with increasing unsaturation, and it is important for estimating the fuel consumption, the greater the calorific value the lower the fuel consumption. Calorific value is usually determined experimentally by a bomb calorimeter, but a theoretical value can be calculated, according to the volumetric concentration of each component in the blend, from the following equation:

$$CV = \sum_i CViXi \quad (2)$$

where CVi is the calorific value of each component and *Xi* is the percentage of each component in the blend [37].

### 2.2.1. Viscosity Measurements

The kinematic viscosity measurements were performed in an Ostwald-Cannon-Fenske capillary viscometer (Proton Routine Viscometer 33200, size 150) at 40 °C, by determining the flow time (t), expressed in seconds, required for a certain volume of liquid to pass under gravity between two marked points on the instrument, placed in an upright position. The kinematic viscosity (υ) expressed in centistokes (cSt) is obtained from the equation $\upsilon = C \cdot t$, where C is the calibration constant of the measurement system, supplied by the manufacturer (0.037150 ($mm^2/s$)/s = cSt at 40 °C) [28,33]. The procedure for the determination of the kinematic viscosity meets with the specifications established by the European standard (EN 590 ISO 3104). All the viscosities values reported here are the media of three determinations.

### 2.2.2. Determination of Pour Point and Cloud Point

Cloud Point (EN 23015 and ASTM D-2500) and Pour Point (ASTM D-97) were determined according to specifications required by standard methods. Firstly, the double or triple blends, of different compositions, were introduced in a flat-bottomed glass tube. The tube was tightly closed with the help of a cork carrying a thermometer with a temperature measuring in the range of −36 to 120 °C. The tube was introduced in a digitally controlled temperature refrigerator for twenty-four hours; the tubes were brought out from time to time and checked until the oil did not show any movement when the tube was horizontally tilted for 5 s. After this time, the loss of transparency of the solutions is evaluated. The appearance of turbidity in the samples is indicative that the CP temperature has been reached. After a progressive decrease in temperature, the samples are kept under observation until they stop flowing (PP) [31,32]. All values are the media of duplicate determinations.

### 2.2.3. Mechanical and Environmental Characterization of a Diesel Engine Electric Generator Fuelled with Different Biofuel Double and Triple Blends

Following the experimental methodology previously described [32], the energy performance and pollutant emissions generated in a C.I. diesel engine, has been carried out, working at a rate of 3000 rpm coupled to an AYERBE AY4000MN electric generator with a power of 5 KVA 230 V, for the generation of electricity, operating under different degrees of demand for electrical power. This is achieved by connecting heating plates of 1000 watts each (Figure 1a). The diesel engine operated at a constant rate of rotation of the crankshaft and torque so that the different values of electrical power obtained are an exact consequence of the mechanical power obtained after the combustion of the corresponding biofuel. Tests were carried out by providing to the engine double and triple blends of different biofuels. The electrical power generated can be easily determined from the product of the potential difference (or voltage) and the electric current intensity (or amperage), Equation 3, both obtained utilizing a voltmeter-ammeter, Figure 1b:

$$\text{Electrical power generated (watts)} = \text{voltage (volts)} \times \text{amperage (amps)} \quad (3)$$

The consumption of the diesel engine, fueled with the different biofuels studied, was calculated estimating the speed of consumption of the engine when it operates under a determined demand of electric power (1, 3 or 5 kW). Thus, the operation times are achieved by operating under the same fuel volume (0.5 L).

The contamination degree is evaluated from the opacity of the smoke generated in the combustion process, which is measured by smoke opacity meters. The smoke opacity meters are instruments capable of measuring the optical properties of diesel exhaust. These instruments have been designed to quantify the visible black smoke emission making use of physical phenomena like the extinction of a light beam by scattering and absorption. There are two groups of instruments: opacity meters, which evaluate smoke in the exhaust gas, and smoke number meters, which optically evaluate soot collected on paper filters. The density gauge is a handheld instrument for determining the filter smoke number (FSN), the Bosch number, and the soot concentration of diesel engines. This instrument is composed of an optical sensor (photodiode) and a differential pressure sensor. The photodiode calculates the paper blackening based on the reflected light intensity by a white LED. The more soot is deposited on the filter paper, the less light is reflected. The probe volume determined by the differential pressure sensor is used to calculate the probe volume under reference conditions with the input height and the temperature measured by the instrument. This probe volume and the measured paper blackening are then used to calculate the FSN (filter smoke number), soot concentration (mg/m$^3$) or Bosch smoke number. Herein, the exhaust emissions were measured by a Bosch smoke meter or opacimeter-type smoke tester TESTO 338 density gauge, following the EU Directive 2004/108/EC, at the operating conditions previously reported [32], Figure 1c. The Bosch number is a standardized unit which is calculated from the level of soot on the paper (effective filter loading) [48]. The instrument evaluates smoke density on a scale from 0 to 2.5, where the value 0 represents total clarity on the paper and 2.5 is the value corresponding to 100% cloudy, as established by ASTM D 2156-94, Standard Test Method for Smoke Density in Flue Gases from Burning Distillate Fuels.

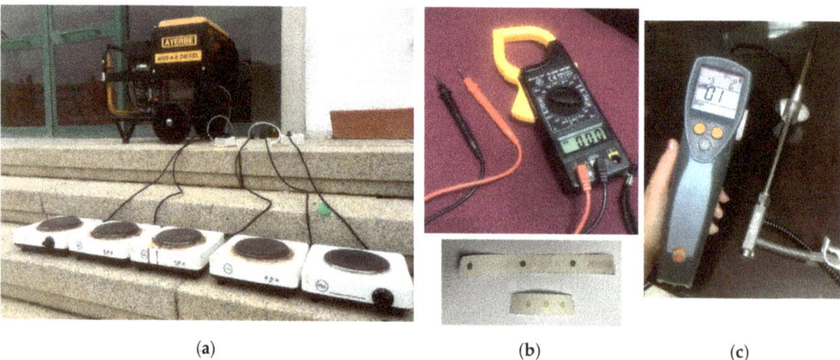


(**a**) (**b**) (**c**)

**Figure 1.** (**a**) Electrogenerator AYERBE AY4000MN, 5 KVA, 230 V connected to heating plates of 1000 watts of power each (**b**) voltmeter-ammeter devise; (**c**) TESTO 338 smoke density tester.

The data here compiled are the media of three repeated measures, attaining an experimental error lower than 9%. The results obtained with the biofuels evaluated were compared with the measurements obtained when conventional diesel was fueled.

## 3. Results and Discussion

### 3.1. Properties of DEE/Oil Double Blends, and Diesel/DEE/Oil Triple Blends

The high viscosity values that the vegetable oils exhibit, 10–20 times greater than fossil diesel fuel, prevents its use like biofuels in conventional diesel engines. Most of them have viscosity values in the range of 30–45 cSt, concretely, castor oil has a much higher value, 226.2 cSt, very superior to values required by European standard EN 590 ISO 3104. Therefore, to achieve adequate viscosity values, the proportion in which the oils must be mixed with DEE has been investigated.

The viscosity values of the DEE/SVOs double blends are shown in Figure 2. As expected, an increase in the DEE content in the blends contributes to decreasing the viscosity values. It is interesting to confirm how DEE can promote very strong action on castor oil, in comparison with sunflower oil, in terms of reducing the viscosity of their blends, since castor oil has initially a much higher viscosity than sunflower oil. As can be observed, only 20% of DEE reduces considerably the kinematic viscosity of both oils, from 226.2 to 17.1 cSt in the case of castor oil and from 37.8 to 13.7 cSt when DEE is added to sunflower oil. For both oils, the viscosity values required by UNE EN 14214 ISO 3104, in the range from 2.0 to 4.5 $mm^2/s$, were achieved with an analogous DEE/vegetable oil ratio (45/55).

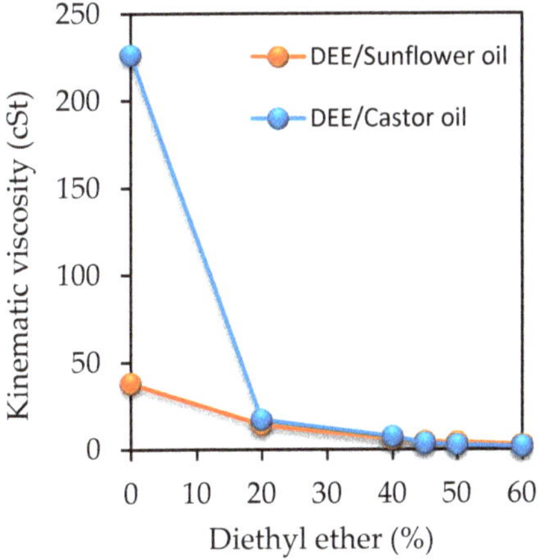


**Figure 2.** Kinematic viscosity values at 40 °C of different blends of DEE/SVOs.

According to the previous study of the kinematic viscosity in double blends, the optimum and most favorable blend ratio was found to be DEE/SVO 45/55 (% by volume), independently of the oil employed. Hence, different proportions of biofuel containing 45% of DEE were mixed with fossil diesel to obtain the diesel/DEE/oil triple blends. The kinematic viscosity, cloud point, pour point, flash point and calorific values of the investigated triple blends are collected in Tables 2 and 3. A higher amount of biofuel in the blend, from B0 to B100, generates higher viscosity values, as expected since the viscosity of the added biofuels is slightly higher than that of diesel (3.20 cSt). Kinematic viscosity values are in the range of 3.20–4.25 cSt, so these biofuels present suitable values for being employed in diesel engines, complying with the European regulations EN 590, which establishes that viscosity at 40 °C must be in the range of 2.0–4.5 cSt.

**Table 2.** Physicochemical properties (kinematic viscosity at 40 °C, cloud point, pour point, flash point, and calorific values) of diesel/DEE/sunflower oil triple blends, obtained by adding different proportions of fossil diesel to the DEE/sunflower oil double blend containing 45% diethyl ether. All values are calculated as the average of three measurements.

| | **Diesel/DEE/Sunflower Oil Blend** | | | | | |
|---|---|---|---|---|---|---|
| Nomenclature | **B0** | **B20** | **B40** | **B60** | **B80** | **B100** |
| (% renewable) | 100/0/0 | 80/9/11 | 60/18/22 | 40/27/33 | 20/36/44 | 0/45/55 |
| Kinematic Viscosity (cSt) | 3.20 ± 0.01 | 3.21 ± 0.01 | 3.26 ± 0.02 | 3.32 ± 0.02 | 3.52 ± 0.02 | 4.25 ± 0.03 |
| Cloud point (°C) | −6.0 ± 1 | −15.0 ± 1 | −15.3 ± 1 | −15.0 ± 1 | −10.6 ± 1 | −15.5 ± 1 |
| Pour point (°C) | −16.0 ± 1 | −21.6 ± 1 | −22.5 ± 1 | −21.9 ± 1 | −20.5 ± 1 | −21.0 ± 1 |
| Flash point (°C) * | 66.0 | 78.4 | 90.9 | 103.3 | 115.8 | 128.2 |
| Calorific value (MJ/kg) * | 42.8 | 41.6 | 40.5 | 39.3 | 38.1 | 36.9 |

* The flash point values and calorific values were calculated by using Equations 1 and 2, respectively.

**Table 3.** Physicochemical properties (kinematic viscosity at 40 °C, cloud point, pour point, flash point, and calorific values) of diesel/DEE/castor oil triple blends, obtained by adding different proportions of fossil diesel to the DEE/castor oil double blend containing 45% diethyl ether. All values are calculated as the average of three measurements.

| | **Diesel/DEE/Castor Oil Blend** | | | | | |
|---|---|---|---|---|---|---|
| Nomenclature | **B0** | **B20** | **B40** | **B60** | **B80** | **B100** |
| (% renewable) | 100/0/0 | 80/9/11 | 60/18/22 | 40/27/33 | 20/36/44 | 0/45/55 |
| Kinematic Viscosity (cSt) | 3.20 ± 0.01 | 3.21 ± 0.01 | 3.24 ± 0.01 | 3.32 ± 0.01 | 3.37 ± 0.02 | 3.48 ± 0.02 |
| Cloud point (°C) | −6.0 ± 1 | −16.0 ± 1 | −16.8 ± 1 | −15.2 ± 1 | −14.2± 1 | −16.3 ± 1 |
| Pour point (°C) | −16.0 ± 1 | −22.3 ± 1 | −22.8 ± 1 | −21.5 ± 1 | −20.0 ± 1 | −22.0 ± 1 |
| Flash point (°C) * | 66.0 | 79.3 | 92.6 | 106.0 | 119.3 | 132.6 |
| Calorific value (MJ/kg) * | 42.8 | 41.4 | 39.9 | 38.5 | 37.1 | 35.7 |

* The flash point values and calorific values were calculated by using Equations 1 and 2, respectively.

Regarding the cold flow properties, the DEE has an important effect in the cloud and pour point of the triple blends, independently of the vegetable oil employed. In fact, it is found that only 9% of DEE (B20 blend) is enough to reduce the PP value 6 °C (from −6 to −15 °C) and the CP value 10 °C (from −6 to −16 °C). The best CP and PP values were obtained adding 18% of DEE, with 22% of castor oil and 60% of diesel, resulting in a CP of −16.8 °C and a PP of −22.8 °C. That is, the DEE promotes a significant improvement, with regard to fossil diesel, on the cold flow properties in all the blends studied, maintaining the appropriate viscosity to be used as a biofuel in conventional diesel engines. Additionally, the use of DEE overcomes one of the major challenges when using biodiesel as an alternative to fossil diesel in current engines, which is its poor cold flow properties.

Tables 2 and 3 show the calorific values of triple blends containing sunflower and castor oil, respectively. The calorific values decreased as the percentage of diethyl ether in the blend increased. As can be observed, there is no notable difference in the respective values for blends with either sunflower or castor oil, since these oils have similar calorific power. The results show that the B20 triple blends exhibited the highest calorific value, 41.6 MJ/kg in the case of sunflower oil and 41.4 MJ/kg for castor oil. In addition, an inverse correlation between calorific value and kinematic viscosity can be observed, since the calorific value of the biofuels increases as the kinematic viscosity decreases.

The results of the flash point of the analyzed mixtures (Tables 2 and 3) show an increment of this value as the DEE/oil ratio is greater, with both sunflower and castor oil. As it can be seen, the incorporation of SVOs in mixtures allows the FPs values to improve, since these oils exhibit a higher FP than both diesel and DEE. The highest FP was presented by B100 blends containing castor oil (132.6 °C) and sunflower oil (128.2 °C). The FP values are situated in the range 78.4–128.2 °C for diesel/DEE/sunflower oil blends, and 79.3–132.6 °C for diesel/DEE/castor oil blends. These values can be compared with FP requirements specified by EN 590 standard, which establishes that the fuel must have a minimum FP of 55 °C. In this case, each of the blends has FPs above the required value, so they are compliant with the requirements. Additionally, these biofuels do not have too high a flash point that they can self–ignite thereby causing no safety problems during handling, storage or transportation, and they are recommended for use in CI engines.

### 3.2. Mechanical Performance of the Diesel Engine

To determine the optimal proportion of DEE that guarantees an adequate engine performance, the biofuel blends characterized in Tables 2 and 3 have been tested in a compression ignition engine. In this study, it was also included for comparative purposes, conventional diesel fuel as a reference. Hence, Figure 3 shows the power generated at different power demanded for by the triple blends employing either sunflower oil (3a) or castor oil (3b). In all cases, the power generated increased as the power supplied to the engine also increased from 1 to 3 kW and then, a stabilization of the power generated occurred, slightly decreasing when the maximum value of power supplied (5 kW) was reached. This behavior was improved using castor oil, where the power generated increased as the power supplied was also increased up to 4 kW.

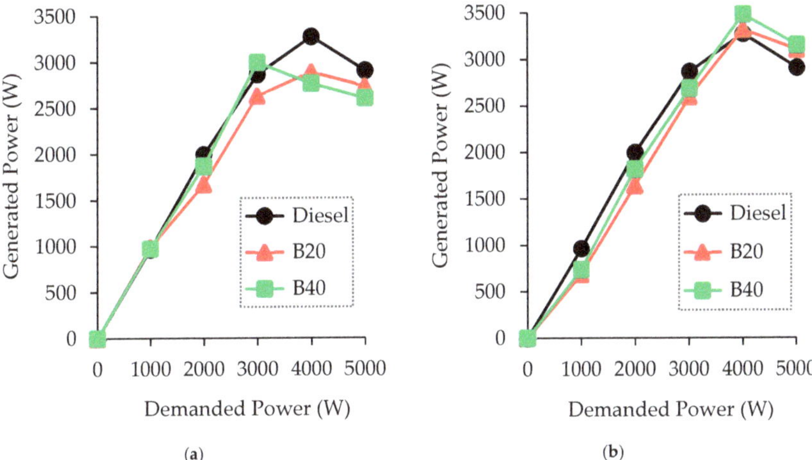


**Figure 3.** Power generated (in Watts) based on the power demanded (in Watts) by the triple blends diesel/DEE/sunflower oil (**a**) or diesel/DEE/castor oil (**b**).

It is noteworthy the excellent results obtained with the B20 and B40 blends, achieving similar (diesel/DEE/sunflower oil) or even higher values (diesel/DEE/castor oil) of power generated than that obtained with fossil diesel at the highest values of demanded power (4 and 5 kW). Independently on the oil employed, a general trend was observed for B20 and B40 blends, where the rise of the DEE content caused the higher power generated. It is important to note that, despite obtaining a very slight increase in the power generated from B20 to B40 blends, when the proportion of DEE in triple mixtures increases above 25%, i.e., B60, B80, and B100 blends, the diesel C.I. engine did not work correctly. This behavior could be associated with the low calorific power of the DEE, confirming the previous results obtained with diesel/ethanol/oil [32].

### 3.3. Smoke Opacity Emissions

Regarding the pollutants emission (Figure 4), the results obtained showed a significant reduction in smoke emissions in diesel engines, compared to those of conventional diesel, especially for B40 blends. This fact was even more evident with castor oil, achieving a reduction of 77% in the pollutant emission at the highest demand employed (5 kW). This important change is mainly due to the contribution of DEE to higher oxygen content in the blend, which improves the combustion and reduces the contamination. So, improved and complete combustion could be the reason for obtaining lower smoke opacity values when the oxygenated additive is added to the blends. Additionally, as can be seen in Figure 4b, a slight but noticeable decrease in smoke emissions is observed when mixtures containing castor oil are employed, compared with the mixtures using sunflower oil in the same proportion (Figure 4a). In previous studies, it has been shown that the presence of unsaturations influences soot formation [49]. So, this may be explained by the differences in the fatty acid composition for the two vegetable oils used, since the linoleic acid present in sunflower oil has a greater number of double bonds than the ricinoleic acid present in castor oil.

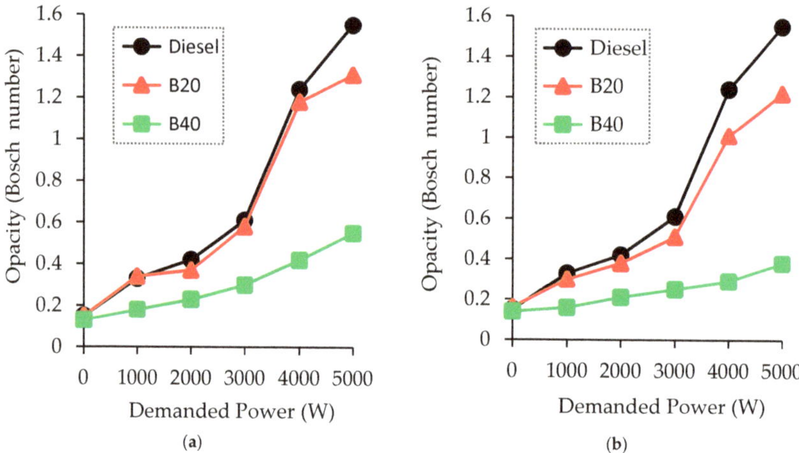


**Figure 4.** Smoke opacity (Bosch number) generated as a function of the power demanded for different triple blends diesel/DEE/sunflower oil (**a**) or diesel/DEE/castor oil (**b**).

It should be noted that independently of the blend tested, from demanded powers of 1000 W onwards, all the blends performed better than fossil diesel in terms of achieving better combustion, which allows a reduction of emissions. The lowest smoke emissions are mainly obtained at medium and high demand (from 2000 W onward). Concretely, a reduction in smoke opacity values of about 66% is achieved with diesel/DEE/sunflower oil blends, whereas the same blends containing castor oil reduce emissions up to 77%.

### 3.4. Fuel Consumption

Additionally, a key parameter in the development of new fuel as an alternative to diesel, is its consumption. Figure 5 shows the consumed volume (in liters per hour) by the engine fueled with the different diesel/DEE/sunflower oil and diesel/DEE/castor oil blends, at different power demands (1, 3, and 5 kW).

As can be seen, at lower power demands (1 kW), the consumption of the blends is always higher than that obtained with the fossil diesel, independently of the oil employed. This is probably due to the initial engine start requires a greater amount of fuel. However, at the higher powers demanded (3 and 5 kW), there have been no noticeable differences in fuel consumption compared to conventional diesel fuel. It can also be seen that mixtures with castor oil (Figure 5b) have slightly higher consumption than sunflower equivalents (Figure 5a). For all fuels tested, the fuel consumption increased as the biofuel ratio (DEE/SVO) is increased in the blends (from B0 to B40). The explanation for this increment could be the reduction in engine power when the concentration of diethyl ether is higher because the energy content in blends is reduced as a consequence of its lower calorific value. In the case of the B40 blend with sunflower oil, fuel consumption was between 3–29% higher than that of diesel, while the biofuel containing castor oil in the same proportion consumed 8–29% more than the diesel case. A reduction of fuel volume up to 9% can be achieved by employing biofuels with 9% of DEE and 11% of vegetable oil, especially when sunflower oil is employed (B20 blend). It was definitive that the tested triple blends B20 and B40 show an excellent ability to eliminate smoke emissions, generate very similar and even higher power values than those with diesel, and a biofuel consumption analogous to diesel.

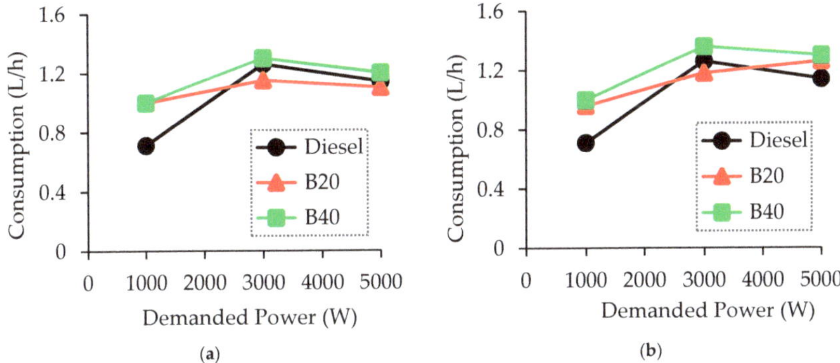


**Figure 5.** Consumption values (L/h) as a function of the power demanded by the engine for the blends diesel/DEE/sunflower oil (**a**) and diesel/DEE/castor oil (**b**).

### 3.5. Comparison with Reported Studies in Literature

The diesel replacement and smoke emissions of the blend, with which the best results are obtained in the present work, have been compared with the results of some of the reported blends as biofuels in the literature, Table 4. It is important to take into account the different parameters in the engine as well as the different fuel loads that have been employed in each study. As can be seen in Table 4 and, to the best of our knowledge, literature about diesel engine fueled with diesel/DEE/oils blends is very recent and limited to a few studies belonging to M. Krishnamoorthi and A. Kumar [36–38,50]. Among these diesel/SVO/DEE triple blends, the greater percentage of substitution has been reached in [38,50] and, also, in the present study (40%). However, with regard to soot emission, a higher reduction of 77%, is achieved in this work, comparing to a reduction from 9.2% to 64.6% obtained with the other triple blends [37,38,50]. The blend containing biodiesel from vegetable oil instead of SVO shows lower diesel replacement and minor reduction of opacity, 22.5% and 8.1%, respectively [40]. In another study, the incorporation of a non−renewable fuel like kerosene only allows a fossil diesel replacement of 13% and the smoke opacities decreased by 6% [51]. There are two trends when double blends containing DEE are employed. On one hand, the blends containing diesel lead to a substitution of diesel up to 24% and to a considerable decrease in emissions [28]. On the other hand, if a renewable fuel like oil derived from waste plastic pyrolysis is used [52], the entire replacement of diesel is attained, at the same time, the pollutant emission is reduced by almost 20% in comparison with the diesel/DEE blend. It has been seen that the triple blends which do not use DEE as an oxygenated additive, for example, the diesel/gasoline/sunflower oil blend [31], allow higher incorporation of SVO than the rest of blends, and generate a similar performance in the diesel engine.

**Table 4.** Comparison of smoke emissions and the percentage of replaced diesel obtained on the blend investigated here and, on several blends reported in the literature.

| Nomenclature | Blend | Diesel Replacement (%) | Smoke Opacity * | Reference |
|---|---|---|---|---|
| B40 | 60% Diesel/22% castor oil/18% DEE | 40 | ↓ 77% | Present study |
| B2 | 60% Diesel/30% aegle marmelos oil/10% DEE | 40 | ↓ 64.6% | [50] |
| B1 | 70% Diesel/20% aegle marmelos oil/10% DEE | 30 | ↓ 9.2% | [37] |
| B2 | 60% Diesel/30% bael oil/10% DEE | 40 | ↓ 64.5% | [38] |

Table 4. *Cont.*

| Nomenclature | Blend | Diesel Replacement (%) | Smoke Opacity * | Reference |
|---|---|---|---|---|
| B20D15 | 68% Diesel/17% cashew Nut Shell Oil/15% DEE | 32 | ↑ 7.3% | [36] |
| B20DEE2.5 | 77.5% Diesel/20% Cottonseed oil biodiesel/2.5%DEE | 22.5 | ↓ 8.1% | [40] |
| DE15K15D | 72.25% Diesel/15% kerosene/12.75% DEE | 13 | ↓ 6% | [51] |
| D+24DEE | 76% Diesel/24% DEE | 24 | ↓ 31% | [28] |
| WD05 | 95% Waste plastic pyrolysis oil/5% DEE | 100 | ↓ 50% | [52] |
| B60 | 40% Diesel/24% gasoline/36% sunflower oil | 36 | ↓ 40% | [31] |

* Smoke opacity reduction with respect to conventional diesel.

## 4. Conclusions

The present study aims to make the process of replacing fossil diesel feasible, without making mechanical changes in the engines of the current car fleet. In this way, not only the pollution produced by petroleum is minimized, but also the dependence of the countries that produce the fuels is reduced. Given the relatively high economic cost of any additional chemical transformation of the vegetable oils, either in biodiesel or in other biofuels of appropriate viscosity values, a novel strategy employing green fuels derived from vegetable oils has been investigated. In this respect, to get high levels of fossil fuel substitution in a technically and economically feasible way, a renewable compound like DEE, which can be obtained from biomass, has been studied as an additive in triple blends with diesel and two differents SVOs on the engine performance and exhaust emissions. DEE has contributed to a reduction of viscosity values in SVOs through its very low viscosity, complying with European regulations to operate as biofuel in current diesel engines. Additionally, all blends comply with EN 590 specifications concerning the flash point.

According to the results obtained, the tests were carried out successfully up to 18% blending of DEE and 22% of SVO with diesel, this means that substitution of 40% of fossil diesel can be achieved by the use of diesel/DEE/SVO triple blends, using either sunflower or castor oil, performing similarly to diesel in terms of power. Furthermore, several advantages of using these blends were obtained. On one hand, DEE addition leads to an improvement in the combustion, since an important reduction of pollutant emissions occurred (up to 77% lower than fossil diesel), which is mainly due to the oxygen content of DEE. Likewise, an enhancement in the cold flow characteristics of the blend in respect to fossil diesel was also attained, which means the diesel engines can run at lower temperatures. On the other hand, the increment of the flash point in blends with respect to fossil diesel, makes these biofuels safer for handling, storage, and transportation. All the abovementioned shows the competitive value of diethyl ether as an oxygenated additive in comparison to other natural products studied of greater economic cost. Hence, the addition of this oxygenated additive up to 18% (by volume) is a promising way for using diesel/vegetable oil blends efficiently in diesel engines without any modifications in the engine. It can be concluded that this research proffers a practical and economically viable alternative to the chemical production of biofuels.

**Author Contributions:** This research article is part of the doctoral thesis of L.A.-D., directed by R.E. and F.M.B., who in general way conceived and designed the experiments. D.L., C.L., J.H.-C., A.P., and A.A.R., made substantive intellectual contributions to this study. L.A.-D. performed the experiments and wrote the paper. R.E., D.L. and F.M.B. supervised the manuscript. All authors have read and agreed to the published version of the manuscript.

**Funding:** This research received no external funding.

**Acknowledgments:** The authors are grateful for the funding received from the Spanish MINECO through the project ENE2016–81013–R (AEI / FEDER, UE), the Junta de Andalucía and FEDER (P18–RT–4822).

**Conflicts of Interest:** The authors declare no conflict of interest.

## Nomenclature

| | |
|---|---|
| ASTM | American society for testing and materials |
| B0 | 100% diesel |
| B20 | 80% diesel + 9% DEE + 11% oil |
| B40 | 60% diesel + 18% DEE + 22% oil |
| B60 | 40% diesel + 27% DEE + 33% oil |
| B80 | 20% diesel + 36% DEE + 44% oil |
| B100 | 45% DEE + 55% oil |
| CI | Compression ignition |
| CN | Cetane number |
| CP | Cloud point |
| cSt | Centistokes |
| CV | Calorific Value |
| DEE | Diethyl ether |
| FAME | Fatty acids methyl esters |
| FP | Flash point |
| FSN | Filter smoke number |
| ISO | International Standards Organization. |
| LVLC | Low viscous and low cetane |
| PP | Pour point |
| rpm | Round per minute ($min^{-1}$) |
| SVO | straight vegetable oils |
| $T_{FP}$ | Flash point temperature |
| W | Watts |
| Symbols | |
| $\upsilon$ | Viscosity (centistokes) |
| t | Flow time (s) |
| C | Calibration constant ($mm^2/s$)/s |

## References


1. Oberthür, S. Where to go from Paris? The European Union in climate geopolitics. *Glob. Aff.* **2016**, *2*, 119–130. [CrossRef]
2. Delucchi, M.A.; Jacobson, M.Z. Meeting the world's energy needs entirely with wind, water, and solar power. *Bull. At. Sci.* **2013**, *69*, 30–40. [CrossRef]
3. Arutyunov, V.S.; Lisichkin, G.V. Energy resources of the 21st century: Problems and forecasts. Can renewable energy sources replace fossil fuels? *Russ. Chem. Rev.* **2017**, *86*, 777. [CrossRef]
4. Sen, B.; Ercan, T.; Tatari, O. Does a battery-electric truck make a difference? Life cycle emissions, costs, and externality analysis of alternative fuel-powered Class 8 heavy-duty trucks in the United States. *J. Clean. Prod.* **2017**, *141*, 110–121. [CrossRef]
5. Noh, H.M.; Benito, A.; Alonso, G. Study of the current incentive rules and mechanisms to promote biofuel use in the EU and their possible application to the civil aviation sector. *Transp. Res. Part D Transp. Environ.* **2016**, *46*, 298–316. [CrossRef]
6. Chu, S.; Majumdar, A. Opportunities and challenges for a sustainable energy future. *Nature* **2012**, *488*, 294. [CrossRef]
7. Lopes, M.; Serrano, L.; Ribeiro, I.; Cascão, P.; Pires, N.; Rafael, S.; Tarelho, L.; Monteiro, A.; Nunes, T.; Evtyugina, M. Emissions characterization from EURO 5 diesel/biodiesel passenger car operating under the new European driving cycle. *Atmos. Environ.* **2014**, *84*, 339–348. [CrossRef]
8. Gardy, J.; Nourafkan, E.; Osatiashtiani, A.; Lee, A.F.; Wilson, K.; Hassanpour, A.; Lai, X. A core-shell SO4/Mg-Al-Fe3O4 catalyst for biodiesel production. *Appl. Catal. B Environ.* **2019**, *259*, 118093. [CrossRef]
9. Gardy, J.; Rehan, M.; Hassanpour, A.; Lai, X.; Nizami, A.-S. Advances in nano-catalysts based biodiesel production from non-food feedstocks. *J. Environ. Manag.* **2019**, *249*, 109316. [CrossRef]
10. Huang, D.; Zhou, H.; Lin, L. Biodiesel: An alternative to conventional fuel. *Energy Procedia* **2012**, *16*, 1874–1885. [CrossRef]

11. Estevez, R.; Aguado-Deblas, L.; Bautista, F.M.; Luna, D.; Luna, C.; Calero, J.; Posadillo, A.; Romero, A.A. Biodiesel at the Crossroads: A Critical Review. *Catalysts* **2019**, *9*, 1033. [CrossRef]
12. Kijenski, J.; Lipkowski, A.; Walisiewicz-Niedbalska, W.; Gwardiak, H.; Rozyczki, K.; Pawlak, I. A Biofuel for Compression-Ignition Engines and a Method for Preparing the Biofuel. European Patent EP1580255, 28 September 2005.
13. Dhawan, M.S.; Yadav, G.D. Insight into a catalytic process for simultaneous production of biodiesel and glycerol carbonate from triglycerides. *Catal. Today* **2018**, *309*, 161–171. [CrossRef]
14. Kim, K.H.; Lee, E.Y. Environmentally-benign dimethyl carbonate-mediated production of chemicals and biofuels from renewable bio-oil. *Energies* **2017**, *10*, 1790. [CrossRef]
15. Luna, D.; Bautista, F.M.; Caballero, V.; Campelo, J.M.; Marinas, J.M.; Romero, A.A. Romero Method for Producing Biodiesel Using Porcine Pancreatic Lipase as an Enzymatic Catalyst. European Patent EP 2 050 823 A1, 22 April 2009.
16. Luna, C.; Luna, D.; Bautista, F.M.; Estevez, R.; Calero, J.; Posadillo, A.; Romero, A.A.; Sancho, E.D. Application of Enzymatic Extracts from a CALB Standard Strain as Biocatalyst within the Context of Conventional Biodiesel Production Optimization. *Molecules* **2017**, *22*, 2025. [CrossRef]
17. Calero, J.; Luna, D.; Sancho, E.D.; Luna, C.; Bautista, F.M.; Romero, A.A.; Posadillo, A.; Berbel, J.; Verdugo-Escamilla, C. An overview on glycerol-free processes for the production of renewable liquid biofuels, applicable in diesel engines. *Renew. Sustain. Energy Rev.* **2015**, *42*, 1437–1452. [CrossRef]
18. Abdulkareem-Alsultan, G.; Asikin-Mijan, N.; Lee, H.; Rashid, U.; Islam, A.; Taufiq-Yap, Y. A Review on Thermal Conversion of Plant Oil (Edible and Inedible) into Green Fuel Using Carbon-Based Nanocatalyst. *Catalysts* **2019**, *9*, 350. [CrossRef]
19. Prakash, T.; Geo, V.E.; Martin, L.J.; Nagalingam, B. Evaluation of pine oil blending to improve the combustion of high viscous (castor oil) biofuel compared to castor oil biodiesel in a CI engine. *Heat Mass Transf.* **2019**, *55*, 1491–1501. [CrossRef]
20. Kasiraman, G.; Nagalingam, B.; Balakrishnan, M. Performance, emission and combustion improvements in a direct injection diesel engine using cashew nut shell oil as fuel with camphor oil blending. *Energy* **2012**, *47*, 116–124. [CrossRef]
21. Vallinayagam, R.; Vedharaj, S.; Yang, W.; Roberts, W.L.; Dibble, R.W. Feasibility of using less viscous and lower cetane (LVLC) fuels in a diesel engine: A review. *Renew. Sustain. Energy Rev.* **2015**, *51*, 1166–1190. [CrossRef]
22. Vallinayagam, R.; Vedharaj, S.; Yang, W.; Lee, P.; Chua, K.; Chou, S. Pine oil–biodiesel blends: A double biofuel strategy to completely eliminate the use of diesel in a diesel engine. *Appl. Energy* **2014**, *130*, 466–473. [CrossRef]
23. Panneerselvam, N.; Ramesh, M.; Murugesan, A.; Vijayakumar, C.; Subramaniam, D.; Kumaravel, A. Effect on direct injection naturally aspirated diesel engine characteristics fuelled by pine oil, Ceiba pentandra methyl ester compared with diesel. *Transp. Res. Part D Transp. Environ.* **2016**, *48*, 225–234. [CrossRef]
24. Vallinayagam, R.; Vedharaj, S.; Yang, W.; Lee, P.; Chua, K.; Chou, S. Combustion performance and emission characteristics study of pine oil in a diesel engine. *Energy* **2013**, *57*, 344–351. [CrossRef]
25. Tamilselvan, P.; Nallusamy, N. Performance, combustion and emission characteristics of a compression ignition engine operating on pine oil. *Biofuels* **2015**, *6*, 273–281. [CrossRef]
26. Subramanian, T.; Varuvel, E.G.; Ganapathy, S.; Vedharaj, S.; Vallinayagam, R. Role of fuel additives on reduction of NO X emission from a diesel engine powered by camphor oil biofuel. *Environ. Sci. Pollut. Res.* **2018**, *25*, 15368–15377. [CrossRef]
27. Rakopoulos, D.C.; Rakopoulos, C.D.; Giakoumis, E.G.; Papagiannakis, R.G.; Kyritsis, D.C. Influence of properties of various common bio-fuels on the combustion and emission characteristics of high-speed DI (direct injection) diesel engine: Vegetable oil, bio-diesel, ethanol, n-butanol, diethyl ether. *Energy* **2014**, *73*, 354–366. [CrossRef]
28. Rakopoulos, D.C.; Rakopoulos, C.D.; Papagiannakis, R.G.; Giakoumis, E.G.; Karellas, S.; Kosmadakis, G.M. Combustion and emissions in an HSDI engine running on diesel or vegetable oil base fuel with n-butanol or diethyl ether as a fuel extender. *J. Energy Eng.* **2016**, *142*, 4015006. [CrossRef]
29. Rakopoulos, D.C.; Rakopoulos, C.D.; Kyritsis, D.C. Butanol or DEE blends with either straight vegetable oil or biodiesel excluding fossil fuel: Comparative effects on diesel engine combustion attributes, cyclic variability and regulated emissions trade-off. *Energy* **2016**, *115*, 314–325. [CrossRef]

30. Lakshminarayanan, A.; Olsen, D.B.; Cabot, P.E. Performance and emission evaluation of triglyceride-gasoline blends in agricultural compression ignition engines. *Appl. Eng. Agric.* **2014**, *30*, 523–534.
31. Estevez, R.; Aguado-Deblas, L.; Posadillo, A.; Hurtado, B.; Bautista, F.M.; Hidalgo, J.M.; Luna, C.; Calero, J.; Romero, A.A.; Luna, D. Performance and Emission Quality Assessment in a Diesel Engine of Straight Castor and Sunflower Vegetable Oils, in Diesel/Gasoline/Oil Triple Blends. *Energies* **2019**, *12*, 2181. [CrossRef]
32. Hurtado, B.; Posadillo, A.; Luna, D.; Bautista, F.; Hidalgo, J.; Luna, C.; Calero, J.; Romero, A.; Estevez, R. Synthesis, Performance and Emission Quality Assessment of Ecodiesel from Castor Oil in Diesel/Biofuel/Alcohol Triple Blends in a Diesel Engine. *Catalysts* **2019**, *9*, 40. [CrossRef]
33. Bailey, B.; Eberhardt, J.; Goguen, S.; Erwin, J. Diethyl ether (DEE) as a renewable diesel fuel. *SAE Trans.* **1997**, *106*, 1578–1584.
34. Kumar, M.V.; Babu, A.V.; Kumar, P.R. The impacts on combustion, performance and emissions of biodiesel by using additives in direct injection diesel engine. *Alex. Eng. J.* **2018**, *57*, 509–516. [CrossRef]
35. Srikanth, H.; Venkatesh, J.; Godiganur, S.; Manne, B. Acetone and Diethyl ether: Improve cold flow properties of Dairy Washed Milkscum biodiesel. *Renew. Energy* **2019**, *130*, 446–451. [CrossRef]
36. Kumar, A.; Rajan, K.; Rajaram Naraynan, M.; Senthil Kumar, K. Performance and emission characteristics of a DI diesel engine fuelled with Cashew Nut Shell Oil (CNSO)-diesel blends with Diethyl ether as additive. *Appl. Mech. Mater.* **2015**, *787*, 746–750. [CrossRef]
37. Krishnamoorthi, M.; Malayalamurthi, R. Experimental investigation on performance, emission behavior and exergy analysis of a variable compression ratio engine fueled with diesel-aegle marmelos oil-diethyl ether blends. *Energy* **2017**, *128*, 312–328. [CrossRef]
38. Krishnamoorthi, M.; Malayalamurthi, R. Availability analysis, performance, combustion and emission behavior of bael oil-diesel-diethyl ether blends in a variable compression ratio diesel engine. *Renew. Energy* **2018**, *119*, 235–252. [CrossRef]
39. Ibrahim, A. An experimental study on using diethyl ether in a diesel engine operated with diesel-biodiesel fuel blend. *Eng. Sci. Technol. Int. J.* **2018**, *21*, 1024–1033. [CrossRef]
40. Yesilyurt, M.K.; Aydin, M. Experimental investigation on the performance, combustion and exhaust emission characteristics of a compression-ignition engine fueled with cottonseed oil biodiesel/diethyl ether/diesel fuel blends. *Energy Convers. Manag.* **2020**, *205*, 112355. [CrossRef]
41. Keera, S.; El Sabagh, S.; Taman, A. Castor oil biodiesel production and optimization. *Egypt. J. Pet.* **2018**, *27*, 979–984. [CrossRef]
42. Ramadhas, A.; Jayaraj, S.; Muraleedharan, C. Use of vegetable oils as IC engine Fuels—A review. *Renew. Energy* **2004**, *29*, 727–742. [CrossRef]
43. Ramadhas, A.S.; Jayaraj, S.; Muraleedharan, C. Biodiesel production from high FFA rubber seed oil. *Fuel* **2005**, *84*, 335–340. [CrossRef]
44. Kwangdinata, R.; Raya, I.; Zakir, M. Production of biodiesel from lipid of phytoplankton Chaetoceros calcitrans through ultrasonic method. *Sci. World J.* **2014**, *2014*, 231361. [CrossRef]
45. Dwivedi, G.; Verma, P.; Sharma, M. Impact of oil and biodiesel on engine operation in cold climatic condition. *J. Mater. Environ. Sci.* **2016**, *7*, 4540–4555.
46. Dwivedi, G.; Sharma, M. Impact of cold flow properties of biodiesel on engine performance. *Renew. Sustain. Energy Rev.* **2014**, *31*, 650–656. [CrossRef]
47. Mejia, J.; Salgado, N.; Orrego, C. Effect of blends of Diesel and Palm-Castor biodiesels on viscosity, cloud point and flash point. *Ind. Crop. Prod.* **2013**, *43*, 791–797. [CrossRef]
48. Sendzikiene, E.; Makareviciene, V.; Janulis, P. Influence of Composition of Fatty Acid Methyl Esters on Smoke Opacity and Amount of Polycyclic Aromatic Hydrocarbons in Engine Emissions. *Pol. J. Environ. Stud.* **2007**, *16*, 259–265.
49. Glassman, I. Soot formation in combustion processes. In *Symposium (International) on Combustion*; Elsevier: Amsterdam, The Netherlands, 1989; Volume 22, pp. 295–311. [CrossRef]
50. Krishnamoorthi, M.; Malayalamurthi, R. Experimental investigation on the availability, performance, combustion and emission distinctiveness of bael oil/diesel/diethyl ether blends powered in a variable compression ratio diesel engine. *Heat Mass Transf.* **2018**, *54*, 2023–2044. [CrossRef]
51. Patil, K.; Thipse, S. Experimental investigation of CI engine combustion, performance and emissions in DEE–kerosene–diesel blends of high DEE concentration. *Energy Convers. Manag.* **2015**, *89*, 396–408. [CrossRef]

52. Devaraj, J.; Robinson, Y.; Ganapathi, P. Experimental investigation of performance, emission and combustion characteristics of waste plastic pyrolysis oil blended with diethyl ether used as fuel for diesel engine. *Energy* **2015**, *85*, 304–309. [CrossRef]

*Article*

# High Vacuum Fractional Distillation (HVFD) Approach for Quality and Performance Improvement of *Azadirachta indica* Biodiesel

**Bazgha Ijaz [1], Muhammad Asif Hanif [1,*], Umer Rashid [2,*], Muhammad Zubair [3], Zahid Mushtaq [4], Haq Nawaz [1], Thomas Shean Yaw Choong [5] and Imededdine Arbi Nehdi [6,7]**

1 Department of Chemistry, University of Agriculture, Faisalabad 38040, Pakistan; bazgha2k14@gmail.com (B.I.); haqchemist@yahoo.com (H.N.)
2 Institute of Advanced Technology, University Putra Malaysia, UPM Serdang, Selangor 43400, Malaysia
3 Department of Chemistry, University of Gujrat, Gujrat 50700, Pakistan; muhammad.zubair@uog.edu.pk
4 Department of Biochemistry, University of Agriculture, Faisalabad 38040, Pakistan; zahidbiochemist@yahoo.com
5 Department of Chemical and Environmental Engineering, University Putra Malaysia, UPM Serdang, Selangor 43400, Malaysia; csthomas@upm.edu.my
6 Chemistry Department, College of Science, King Saud University, Riyadh 1145, Saudi Arabia; imed12002@gmail.com
7 Chemistry Department, Preparatory Institute for Engineering Studies of EI Manar, Tunis El Manar University, P.O. Box. 244, Tunis 2092, Tunisia
* Correspondence: drmuhammadasifhanif@gmail.com (M.A.H.); umer.rashid@upm.edu.my (U.R.); Tel.: +92-3338362781 (M.A.H.); +60-3-9769-7393 (U.R.)

Received: 19 March 2020; Accepted: 23 April 2020; Published: 3 June 2020

**Abstract:** Biodiesel offers an advantage only if it can be used as a direct replacement for ordinary diesel. There are many reasons to promote biodiesel. However, biodiesel cannot get wide acceptance until its drawbacks have been overcome including poor low temperature flow properties, variation in the quality of biodiesel produced from different feedstocks and fuel filter blocking. In the present study, a much cheaper and simpler method called high vacuum fractional distillation (HVFD) has been used as an alternative to produce high-quality refined biodiesel and to improve on the abovementioned drawbacks of biodiesel. The results of the present study showed that none of biodiesel sample produced from crude *Azadirachta indica* (neem) oil met standard biodiesel cetane number requirements. The high vacuum fractional distillation (HVFD) process improved the cetane number of produced biodiesels which ranged from 44–87.3. Similarly, biodiesel produced from fractionated *Azadirachta indica* oil has shown lower iodine values (91.2) and much better cloud (−2.6 °C) and pour point (−4.9 °C) than pure *Azadirachta indica* oil. In conclusion, the crude oil needs to be vacuum fractioned for superior biodiesel production for direct utilization in engine and consistent quality production.

**Keywords:** biodiesel; vacuum fractionation; transesterification; fuel; fatty acids composition

## 1. Introduction

Biodiesel is getting the attention of policy makers to overcome serious concerns about climate change and maintain the security of the energy supply. Biodiesel has several advantages, such as being non-toxic, sustainable, non-explosive, environmentally friendly and biodegradable, which lessens toxic emissions and productions when used in a diesel engine [1–8]. Many observers consider biodiesel to be the one of the feasible options for the substitution of fossil diesel in the transport sector [9–19]. Increasing biodiesel supplies helps to reduce fuel imports and to cut down the emission of greenhouse gases (GHG). There is a downside of biodiesel as well. The fuel performance in in the engine depends upon the derived cetane number, total acid number, viscosity, and oxidative stability, etc. Biodiesels usually have higher viscosity than petroleum diesel. The engine will not operate well when the viscosity of the fuel is too high. The derived cetane number (CN) is "the measure of the ignition delay from the time the fuel injected and the start of combustion". The higher the CN, the shorter the ignition delay and, therefore, the better the quality of the fuel [5,8,20–22].

Most importantly, food crop production for biodiesel often requires massive acreage and may lead to land-use conflicts in food production. Despite these using non-edible oil seed crops such as *Pongamia glabra, Jatropha curcas* and *Azadirachta indica* prove best suitable for the biodiesel synthesis. Several other disadvantages include higher cost, poor low temperature, flow properties, variation in ignition quality of biodiesel produced from different feedstocks and clogging in engine etc. Biodiesel is unusable in cold areas. This is one of the major drawbacks of biodiesel use. If it gets below −1 °C, it will be solidifying in the engine and fuel tank. Usually, the temperature of congelation is relatively high for biodiesel. However, biodiesel can still be used in winter, if it is mixed with some sort of winterized diesel to remain a liquid. Biodiesel cold-flow properties are characterized by three temperature measures: cloud point (CP), "the temperature at which the fuel shows a haze from the formation of crystals"; cold filter plugging point (CFPP), "the temperature at which the crystals formed will cause the plugging of the filters"; and pour point (PP), "the lowest temperature at which the liquid will flow" [23].

The crude oils usually contains moisture, gums (lecithins), solids (insoluble), waxes, free fatty acids (FFA), and compounds of Na, K, Mg, Ca and other metals, which must be removed to make high-quality biodiesel more efficient and stable against rancidity upon storage. A series of steps are used to remove these impurities, including degumming (to remove gums), neutralizing (to remove FFA), bleaching (to remove color), deodorizing (to remove odor and taste), and dewaxing or winterization (to remove waxes). The alternative methodology to improve quality of biodiesel adopted in this study was fractionation of *Azadirachta indica* seed oil followed by its transesterification for production of biodiesel. The high demand for biodiesel in replacing fossil fuels has driven many researchers to come out with new ideas and inventions. Neem (*Azadirachta indica*) belongs to the Maliaceae family. *Azadirachta indica* seed oil is a non-edible feedstock to produce biodiesel. *Azadirachta indica* seed contains up to 40% lipid contents [24]. The main purpose of the present study was to remove the components from the source oil that deteriorate the quality of the produced biodiesel. Biodiesel quality can be directly related to type and percentage amount of various fatty acids present in the source oil. The presence of fatty acids with short chains or overlong chains could have negative effects on biodiesel fuel standard parameters. Hence, it is possible to control and improve the quality of the produced biodiesel by controlling and maintaining the fatty acid composition of the source oil. In the present study, *Azadirachta indica* seed oil is separated into several fractions using high vacuum fractional distillation (HVFD) to produce biodiesel with superior flow and burning proprieties. The present study is based on the hypotheses that biodiesel produced from a specific oil fraction show more consistent fatty acid composition, low cloud point and low pour point.

## 2. Materials and Methods

### 2.1. Sample Collection and Oil Extraction

The *Azadirachta indica* seeds (10 kg) were collected from the University of Agriculture Faisalabad, Pakistan. A sample cleanup using deionized distilled water (DDW) was carried out to eliminate impurities and dirt. The cleaned samples were dried at 40 °C in an electric oven until a constant weight was attained. The *Azadirachta indica* seed oil extracted using a cold press.

### 2.2. Vacuum Fractional Distillation of Azadirachta Indica Oil

Vacuum fractionation distillation of *Azadirachta indica* oil was carried out to separate out different isolates based on boiling points. Vacuum fractional distillation apparatus consisted of an electric heater (operation range: room temperature to 375 °C), boiling flask (500 $dm^3$), condenser with vacuum adapter, one-stage vacuum pump (10 Pa, power = $\frac{1}{4}$, oil capacity = 250 $dm^3$, Model TW-1A), short-path distillation receiver, cow shaped (with four 50 $dm^3$ flasks). In a single run, 300 $dm^3$ of *Azadirachta indica* oil was separated into fractions under a constant vacuum of −760 mmHg at varied temperatures. A digital thermometer was used to record temperature of vapors of boiling fractions (Figure 1). Table 1 present fractions obtained after the distillation of 650 g of *Azadirachta indica* oil. The oil left after fractionation consist of impurities that severely affect biodiesel properties. This oil can be further processed to produce lubricating liquid, *Azadirachta indica*-based antiseptic soap or in monoacylglycerol production as done previously in the literature [25]. The *Azadirachta indica* oil was distilled using vacuum fractions to isolate fraction F1 (120–150 °C), fraction F2 (180–190 °C), fraction F3 (202–230 °C) and fraction F4 (235–240 °C).

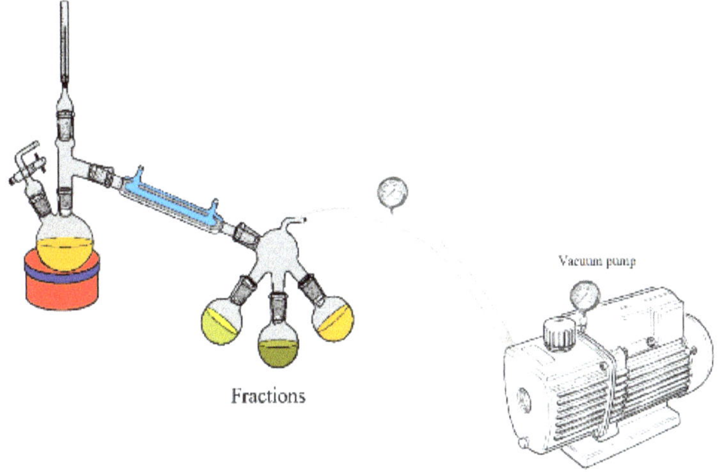


**Figure 1.** High vacuum fractionation distillation (HVFD) setup of *Azadirachta indica* oil.

**Table 1.** *Azadirachta indica* seed oil fractions separated using vacuum fractionation at −760 mmHg.

| Fraction | Temperature Range (°C) | Wt. (g) | % of Total *Azadirachta indica* Oil |
|---|---|---|---|
| F1 | 120–150 | 156.48 | 24.06 ± 0.87 |
| F2 | 180–190 | 69.54 | 10.68 ± 0.67 |
| F3 | 202–230 | 111.35 | 17.12 ± 0.54 |
| F4 | 235–240 | 144.61 | 22.24 ± 0.97 |

### 2.3. Transesterification

Transesterification is a cheaper way of converting the branched, large molecules of oil to smaller molecules [26]. The extracted *Azadirachta indica* oil and its fractions were converted into biodiesel and glycerol (byproduct) using methanol through the catalytic transesterification process. The transesterification process was carried out by base (potassium hydroxide), enzyme (lipase Novozymes) and acid (hydrochloric acid) catalysts. Each catalyst was used at five different concentration levels for the optimization of the catalyst dose required for the transesterification process. The concentration of catalyst was measured in percentage (weight of catalyst/weight of oil). The base catalyst (KOH) was used at five different concentrations levels (0.2%, 0.4 %, 0.6%, 0.8% and 1% of the *Azadirachta indica* oil weight). The mixture of KOH (of respective concentration), *Azadirachta indica* oil or its fractions (10 g) and methanol (3 g) was stirred at 60 °C for 150 min. For the acid-catalyzed transesterification, HCl concentrations were 20%, 40%, 60%, 80% and 100% of oil weight. The used concentration of methanol was 5%. A mixture of oil, methanol and acid was stirred at 75 °C for 270 min. Five concentrations (1%, 2%, 3%, 4% and 5%) of lipase were added to the mixture of oil (10 g) and methanol (50 g) under constant stirring at 100 rpm for 24 h. The water insoluble biodiesel layer and glycerol (water soluble) layer were separated using a 250 $dm^3$ separatory funnel. The biodiesel layer was washed with an excess quantity of warm deionized distilled water (DDW) to remove soap and other water-soluble contaminants. Biodiesel was dried over anhydrous sodium sulfate before further use.

### *2.4. Biodiesel Quality Testing*

The chemical composition of *Azadirachta indica* oil and its fractions was analyzed by gas chromatographic and mass spectrometric (GC-MS) analysis. A Perkin Elmer Clarus 600 GC System was used for the GC-MS analyses fitted with a capillary column (30.0 m–0.25 mm) at the 350-°C optimum temperature. As carrier gas, ultra-high (99.99%) pure helium was used with 0.2 $dm^3$/min constant flow. The transfer line, ion source and injection temperatures were 200, 200 and 220 °C, respectively. There was 70 eV ionization energy in the system for the breakdown of fragments. The data collection was done by mass spectra ranging from 10 to 600 m/z. Then, 0.1 $dm^3$ of the injected sample was used with 50:1 split ratio. The identification of obtained compounds was done by comparison with mass spectrum libraries (Wiley, 9th edition) and retention times.

The physico-chemical properties such as densities (kg/lit), saponification values (mg KOH/g), iodine values, cetane number, cloud and pour points were tested by following standard procedures [2,27]. The cold filter plugging point (CFPP) was determined in accordance with the EN 116 standard procedure for CFPP testing, as specified in the EN 590 standard [28]. Biodiesel pH was measured using a Hanna pH meter (model, HI 8010), and a densitometer (Wilnos LCD 51) was used for density (gram per $dm^3$) determination. The oxidation stability was measured by the Induction Period (IP) according to EN14112 [29].

For iodine value (I.V) determination, 0.05 g of *Azadirachta indica* biodiesel was taken in an iodine flask with a capacity of 250 $dm^3$. Then, 12.5 $dm^3$ of Wijs solution and 10 $dm^3$ of carbon tetrachloride were dissolved in biodiesel. Subsequently, contents of mixture were shaken vigorously and kept in a dark place for almost 30 min followed by the addition of 10 $dm^3$ of potassium iodide solution (15%) and 50 $dm^3$ of DDW. The obtained mixture was titrated against $Na_2S_2O_3.5H_2O$ (0.1 N) until the disappearance of the color of the iodine using starch as an indicator. For the blank sample, a similar procedure was adopted. The following formula (Equation (1)) was used for the determination of the iodine value of biodiesel samples [2]:

$$\text{Iodine value} = \frac{(\text{Blank titration} - \text{Sample titration}) \times \text{Normality of } Na_2S_2O_3.5H_2O \times 12}{\text{Sample weight (g)} \times 100} \quad (1)$$

To determine the saponification value, 10 $dm^3$ KOH (alcoholic) solution and 0.25 g *Azadirachta indica* oil were taken in a round bottom flask of 250 $dm^3$ capacity. A condenser was attached with this flask and heated smoothly till transparent solution was obtained. The appearance of a transparent solution indicates the completion of the saponification reaction. After cooling the reaction mixture to room temperature, a phenolphthalein indicator (2–3 drops) was added to it and the whole mixture was titrated against HCl (0.5 N) until the disappearance of the pink color. An appropriate reagent blank sample was also prepared, and a reading was determined for it. Using the following formula (Equation (2)), the saponification value in mg KOH/g of oil sample was calculated [2,30]:

$$\text{Saponification value (SV)} = \frac{(B - S) \times N \times 56.1}{W} \quad (2)$$

where S is the volume used of titrant for biodiesel sample, B is the volume used of titrant for the blank sample, 56.1 is MW of KOH in mg per mmol, N is the normality of HCl in mmol per $dm^3$ and w shows sample mass (gm).

For the determination of acid value, 1g of *Azadirachta indica* oil, 20 $dm^3$ of ethanol and 2–3 drops of phenolphthalein were added to a 250-$dm^3$ titration flask. This solution was titrated against NaOH (0.1 N) solution until there was a pink appearance. Using the following equation (Equation (3)), free fatty acid contents (oleic acid) were determined [2]:

$$\text{Free fatty acid (FFA)}, \% = \frac{(V \times N \times 282)100}{w} \tag{3}$$

where V stands for volume of titrant sodium hydroxide ($dm^3$), N for Normality of sodium hydroxide (mol/1000 $dm^3$), 282 is the molecular weight of oleic acid in gram per mole. The obtained value was converted into acid value using the following equation (Equation (4)):

$$\text{Acid value (AV)} = 1.989 \times \%\ \text{FFA} \tag{4}$$

The following formula (Equation (5)) was used to determine the Cetane number (CN) of all biodiesel samples:

$$\text{Cetane number (CN)} = 46.3 + \frac{5458}{SV} - 0.225 \times IV \tag{5}$$

## 3. Results and Discussion

### 3.1. Effect of Various Catalysts on the Biodiesel Yield Percentage (%)

The effect of different catalysts (acid, base and enzyme) on the percentage yield of biodiesel is shown in Figure 2a–c. Transesterification using base catalysts is found to be cost effective because of its reusability, wide availability, easy separation from product and longer lifetime [31,32]. In base-catalyzed transesterification of *Azadirachta indica* oil and its fractions, the maximum biodiesel yield of 99.9% was shown by fraction F4. In another study, the biodiesel yield obtained by whole date seed oil using base catalysts was approximately 80% [2]. The optimized catalyst level for pure oil, F1 and F4, was 1% KOH (wt. of catalyst/wt. of oil) and for F2 and F3 was 0.2% KOH (wt. of catalyst/wt. of oil). Thus, the result obtained clearly indicated that, for better biodiesel yields, higher alkaline catalyst doses are required in the case of *Azadirachta indica* oil, F1 and F2. However, fractions F2 and F3 produced the highest quantity of biodiesel at the lowest KOH dose of 0.2% KOH (wt. of catalyst/wt. of oil). This indicates that after fractionation biodiesel could be produced at much lower levels of base catalyst using some specific fractions. This will be helpful in cutting costs on base catalysts and will also be helpful in reducing environmental pollution caused by base catalysts. The maximum biodiesel yields of acid catalyzed transesterification reactions were 99%, 85%, 97.25%, 76.25% and 94.25%, respectively, for *Azadirachta indica* oil, fraction F1, fraction F2, fraction F3 and fraction F4. The most suitable level of HCl for obtaining maximum biodiesel from *Azadirachta indica* oil, F3 and F4 was 80% (wt. of HCl/wt. of oil). However, F1 and F2 have shown maximum biodiesel yields at 100% and 60%, respectively. The acid catalysis reaction rate was much slower than base catalysis. The acid-catalyzed transesterification reaction was completed in 4.5 h as compared to the 1.5 h taken by the base-catalyzed reaction. Transesterification is the replacement of original ester groups into desired esters. The best way for transesterification is base-catalyzed as it is a very facile reaction. The enzyme is inhibited in the presence of methanol. Acidic catalysts have low acid site concentration, low micro porosity, and high cost compared with basic types. Acids catalyzes both esterification and transesterification simultaneously and is insensitive to FFA and water. In a previous study, biodiesel productivities of used vegetable oils by acidic catalysis at 25 and 100 °C, respectively, were as follows: Safflower (84.7% and 94.3%), Soybean (85.9% and 94.2%), Sunflower (83.4% and 95.2%), Canola (80.8%

and 93.7%), Corn (83.2% and 83.3%), Olive (84.3% and 85.3%), Hazelnut (82.5% and 83.4%) and Waste sunflower (84.3% and 90.4%) [33].

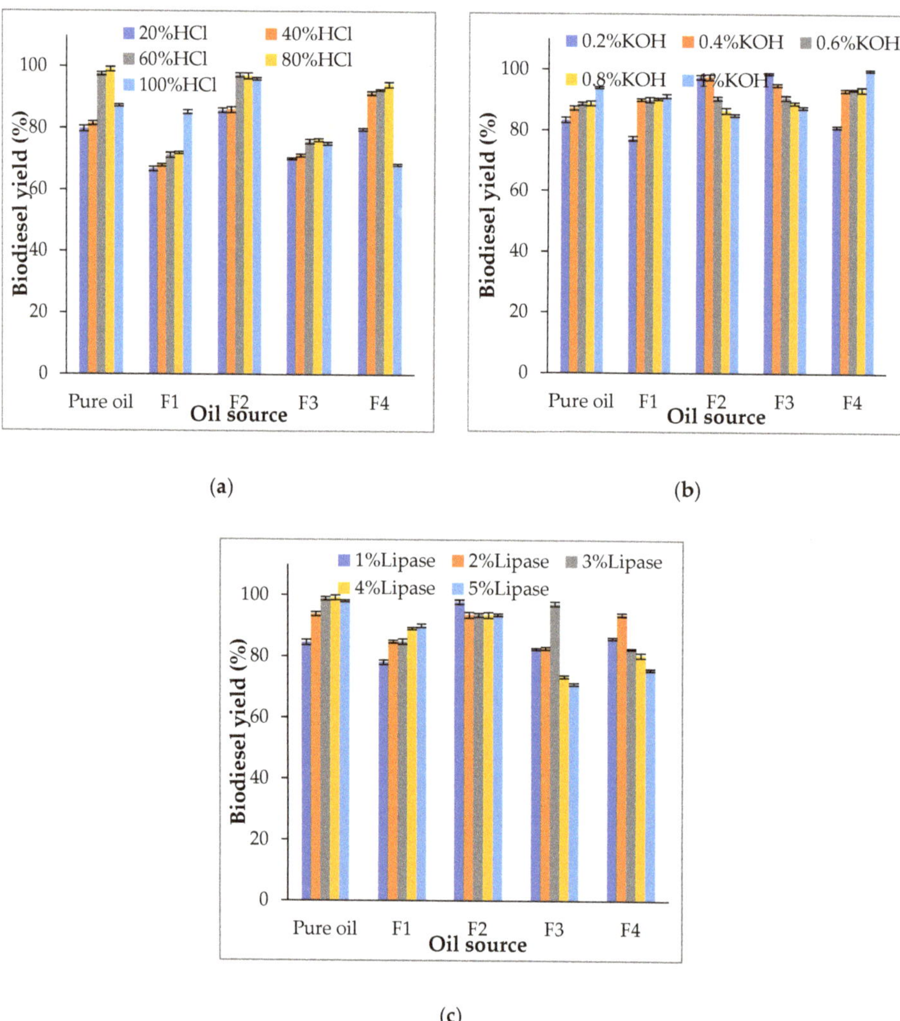


**Figure 2.** (**a**) Biodiesel yield using HCl as the catalyst; (**b**) biodiesel yield using KOH as the catalyst; (**c**) biodiesel yield using lipase enzyme as the catalyst.

Enzyme-catalyzed biodiesel production was completed in 24 h. For the biodiesel synthesis, five different lipase concentrations (1–5%) were used. The results obtained show that *Azadirachta indica* oil and its four fractions have the highest biodiesel yield at different enzyme concentrations. F2, F4, F3, pure oil and F1 produced the maximum quantity of biodiesel at 1%, 2%, 3%, 4% and 5% of lipase, respectively, and the respective maximum yields were 97.75%, 93.75%, 97.25%, 99.1% and 89.75%. The *Azadirachta indica* oil fractionation process yielded distillates with a variable composition as confirmed by gas chromatographic and mass spectrometric analysis. The variance in composition of fractions and *Azadirachta indica* oil was responsible for the variable optimized yield. Enzymes have a three-dimensional structure and are large protein biomolecules that can take part in many chemical reactions. Enzymes are highly specific; they function with only one reactant to produce specific products. Comparing acid, base and enzyme catalyst-optimized yields, it is apparent that the highest biodiesel yield was shown by base-catalyzed transesterification. The biodiesel yields of acid and enzyme-catalyzed reactions were comparable.

### 3.2. Gas Chromatographic and Mass Spectroscopic (GC-MS) Analysis

The identification and quantification of the fatty acids present in biodiesel was conducted by GC-MS. The major fatty acids present in *Azadirachta indica* oil were found to be oleic acid (67%), palmitic acid (14.7%) and lignoceric acid (8.23%). The fuel and physical properties of biodiesel are dependent on chain length, degree of unsaturation, and branching of the chain. In a previous study, the quality of biodiesel produced from shorter fatty acids was superior than longer chain fatty acids [34]. The longer carbon chain length leads to decreased ignition delay. Longer chain fatty acids are present in all oils. However, their percentages in various oils widely differ. Lignoceric acid, or tetracosanoic acid, is the saturated fatty acid with the formula $C_{23}H_{47}COOH$. It is present in almost 8.23% of *Azadirachta indica* oil. Its removal from *Azadirachta indica* oil was necessary to avoid its negative effects on the biodiesel centane number, cloud point, pour point and freeze point. The vacuum fractional distillation process was used to carry out lignoceric acid removal from biodiesel. Lignoceric acid was removed from oil and collected in the first fraction F1. The other major components of F1 were erucic acid (22.4%), oleic acid (20.9%), phthalic acid (13.4%) and linoleic acid (11.7%). Fraction F2 contained oleic acid (C18, 61.3%) as a major fatty acid and all other fatty acids were present at less than 10%. Capric acid (C10) was 43.6% present in the fraction F3. Fraction F4 was found to have highest quantities of oleic acid (71.8%). Linoleic acid (C18) was present 16.91% in F4 (Table 2).

**Table 2.** Fatty acid composition of *Azadirachta indica* oil and its fractions.

| No. | Fatty Acid | Percentage (%) | | | | |
|---|---|---|---|---|---|---|
| | | F-1 | F-2 | F-3 | F-4 | *Azadirachta indica* Oil |
| 1 | Phthalic acid (C6) | 13.4 | 3.9 | 0 | 0 | 2.6 |
| 2 | Caprylic Acid (C8) | 1.45 | 0 | 0 | 0 | 1.44 |
| 3 | Capric acid (C10) | 0 | 0 | 43.6 | 0 | 0 |
| 4 | Dichloroacetic acid, undec-2-enyl ester (C13) | 0 | 0 | 26.0 | 0 | 0 |
| 5 | Palmitic Acid (C16) | 5.3 | 2.32 | 18.5 | 1.3 | 14.7 |
| 6 | Linoleic Acid (C18) | 11.7 | 0 | 0.60 | 16.91 | 1.30 |
| 7 | Oleic Acid (C18) | 20.9 | 61.3 | 5.1 | 71.8 | 67.0 |
| 8 | Stearic Acid (C18) | 2.72 | 4.8 | 0 | 1.7 | 0.87 |
| 9 | Arachidic Acid (C20) | 4.2 | 8.38 | 0 | 3.1 | 1.88 |
| 10 | Behenic Acid (C22) | 5.9 | 8.6 | 0 | 0 | 0.98 |
| 11 | Erucic Acid (C22) | 22.4 | 9.4 | 0 | 1.7 | 0.93 |
| 12 | Lignoceric acid (C24) | 10.6 | 0 | 0 | 0 | 8.23 |

### 3.3. Assessment of Fuel Quality Parameters

The acid value (AV) is defined as "the weight of KOH in mg needed to neutralize the organic acids present in 1 g of fat and it is a measure of the free fatty acids (FFA) present in the fat or oil" [31]. An increment in the amount of FFA in a sample of oil or fat indicates hydrolysis of triglycerides. The acid value (AV) of *Azadirachta indica* oil, F1, F2, F3 and F4 were 1.23, 2.91, 6.61, 8.89 and 3.36, respectively. The results obtained show that there was breakdown of fatty acids from triglyceride chains during the molecular distillation process of fractionation. The observed density of all biodiesel samples is given in Table 3. According to American standards (ASTM), there is no specified range of density of produced biodiesel. However, in European standards (EN), the recommended range of densities of biodiesel should range from 0.86 to 0.90 kg/liter. The density of *Azadirachta indica* oil and fraction biodiesel samples ranged from 0.70–0.91 kg/liter. The density value has great influence on some properties of the biodiesel concerning the engine efficiency including injection timing, injection system and spray properties. Lower density biodiesel has lower engine efficiency and higher density value exhibits greater mass of biodiesel introduced into the injection system [35].

**Table 3.** Densities (kg/dm$^3$) of biodiesel produced from *Azadirachta indica* seed oil and its fractions.

| Catalyst | Conc. of Catalyst (%) | Densities (kg/dm$^3$) | | | | |
|---|---|---|---|---|---|---|
| | | *Azadirachta indica* oil | F-1 | F-2 | F-3 | F-4 |
| HCl | 20 | 0.90 | 0.89 | 0.87 | 0.81 | 0.84 |
| | 40 | 0.89 | 0.86 | 0.87 | 0.76 | 0.83 |
| | 60 | 0.86 | 0.80 | 0.81 | 0.75 | 0.82 |
| | 80 | 0.85 | 0.81 | 0.79 | 0.75 | 0.81 |
| | 100 | 0.84 | 0.80 | 0.78 | 0.74 | 0.81 |
| KOH | 0.2 | 0.86 | 0.85 | 0.82 | 0.82 | 0.90 |
| | 0.4 | 0.84 | 0.84 | 0.80 | 0.81 | 0.90 |
| | 0.6 | 0.83 | 0.82 | 0.78 | 0.78 | 0.88 |
| | 0.8 | 0.82 | 0.80 | 0.75 | 0.78 | 0.76 |
| | 1.0 | 0.80 | 0.79 | 0.75 | 0.77 | 0.70 |
| Lipase | 1 | 0.86 | 0.90 | 0.89 | 0.86 | 0.85 |
| | 2 | 0.80 | 0.90 | 0.89 | 0.76 | 0.84 |
| | 3 | 0.80 | 0.88 | 0.87 | 0.76 | 0.84 |
| | 4 | 0.80 | 0.86 | 0.87 | 0.75 | 0.82 |
| | 5 | 0.79 | 0.84 | 0.81 | 0.75 | 0.75 |

The saponification value or saponification number of *Azadirachta indica* biodiesel synthesized by using several catalyst concentrations is given in the Table 4. The saponification value or number relies on fatty acid concentration present in the biodiesel and the molecular weight. The existence of fatty acids with short chains of alkyl groups in biodiesel is also indicated by the saponification number. Using a KOH catalyst, higher saponification values of *Azadirachta indica* biodiesel were observed. This indicated high volatilities of the synthesized biodiesel. This might be supportive in burning *Azadirachta indica* biodiesel easily and efficiently to avoid the backfire. This happens in diesel engines when unburned fuel meets hot parts that are close enough to the tip, where oxygen is present, to explode spontaneously. However, it is quite rare for this to happen. This property is reflected in the existence of the small chained fatty acids or methyl esters.

**Table 4.** Saponification values (mg KOH/g) of biodiesel from *Azadirachta indica* seed oil and its various fractions.

| Catalyst | Conc. of Catalyst (%) | Saponification Number | | | | |
|---|---|---|---|---|---|---|
| | | *Azadirachta indica* oil | F-1 | F-2 | F-3 | F-4 |
| HCl | 20 | 270.2 ± 4.5 | 202.9 ± 4.1 | 270.2 ± 4.7 | 202.9 ± 4.1 | 270.2 ± 4.7 |
| | 40 | 270.2 ± 4.5 | 214.1 ± 4.2 | 146.8 ± 2.6 | 191.9 ± 2.9 | 158.0 ± 2.5 |
| | 60 | 171.9 ± 2.5 | 158.0 ± 2.6 | 191.7 ± 2.9 | 101.9 ± 2.1 | 158.0 ± 2.5 |
| | 80 | 202.9 ± 4.1 | 214.1 ± 2.1 | 180.5 ± 2.8 | 101.9 ± 2.1 | 202.9 ± 4.2 |
| | 100 | 281.4 ± 4.8 | 124.4 ± 2.2 | 191.7 ± 2.9 | 146.4 ± 2.4 | 109.7 ± 2.9 |
| KOH | 0.2 | 308.5 ± 5.3 | 249.0 ± 4.5 | 270.2 ± 4.7 | 281.4 ± 4.8 | 214.1 ± 4.1 |
| | 0.4 | 341.5 ± 5.5 | 204.9 ± 4.1 | 281.4 ± 4.8 | 276.8 ± 4.7 | 259.0 ± 4.5 |
| | 0.6 | 360.0 ± 5.6 | 156.1 ± 2.5 | 281.4 ± 4.8 | 270.2 ± 4.7 | 136.6 ± 2.3 |
| | 0.8 | 371.2 ± 5.7 | 185.8 ± 2.8 | 191.7 ± 2.9 | 270.2 ± 4.7 | 270.2 ± 4.7 |
| | 1 | 376.1 ± 5.8 | 178.5 ± 2.7 | 214.1 ± 4.2 | 259.5 ± 4.5 | 225.3 ± 4.2 |
| Lipase | 1 | 214.1 ± 4.2 | 158.0 ± 2.5 | 202.9 ± 4.3 | 226.3 ± 4.2 | 124.4 ± 2.2 |
| | 2 | 202.9 ± 4.1 | 180.5 ± 2.8 | 124.4 ± 2.4 | 157 ± 2.5 | 169.2 ± 2.6 |
| | 3 | 191.7 ± 2.9 | 146.8 ± 2.4 | 191.7 ± 2.9 | 124.4 ± 2.2 | 101.9 ± 2.1 |
| | 4 | 191.7 ± 2.9 | 135.6 ± 2.3 | 169.2 ± 2.6 | 134.6 ± 2.3 | 101.9 ±2.1 |
| | 5 | 180.5 ± 2.8 | 158.0 ± 2.5 | 214.1 ± 4.1 | 134.6 ± 2.3 | 124.4 ± 2.2 |

Furthermore, a high saponification number also articulates the presence of carboxylic groups in high amounts in the biodiesel. In the present study, the maximum saponification value was 376 mg

KOH/g of biodiesel sample synthesized by the pure *Azadirachta indica* oil using 1% KOH catalyst (w/w of oil), while the minimum saponification value (136.6 mg KOH/g) of biodiesel sample produced by the F4 fraction. Using the HCl catalyst, a maximum saponification value (281.4) of pure *Azadirachta indica* oil biodiesel was shown using 100% HCl catalyst (w/w of oil) and the minimum value (101.9) for the F3 fraction was observed using 60% and 80% HCl catalyst w/w of oil. Using the lipase enzyme catalyst, a maximum saponification value (226.3) was observed for the F3 fraction using 1% lipase enzyme catalyst (w/w of oil), while the minimum SP value (101.9) was observed for the F4 fraction using 3% and 4% lipase enzyme catalyst (w/w of oil).

The iodine values (IV) of *Azadirachta indica* biodiesel are shown in Table 5. The minimum iodine value (37.2) for the *Azadirachta indica* biodiesel was found for the biodiesel synthesized using 20% HCl catalyst w/w of oil, while the maximum iodine value (218.4) was shown by pure *Azadirachta indica* oil transesterified using KOH 0.8% w/w of oil. Iodine value relies on the unsaturated fatty acid and indicate double bound present in biodiesel sample [36]. According to European standards, iodine values equal to or less than 120 is preferable for biodiesel produced for consumption. However, in American standards, no iodine value is specified. Unsaturation represented by the iodine value reflects the chance of biofuel solidification [37]. The greater the unsaturation, the greater the iodine value will be. High unsaturation in the biofuel may cause deposition as breakage of some weak bonds results in polymerization or leads to epoxide formation in the engine at high temperatures. A careful examination of Table 5 shows that biodiesel produced from *Azadirachta indica* oil had higher iodine values. However, fractionation of *Azadirachta indica* oil into various components was useful for producing better quality biodiesel with lower iodine values. The production of biodiesel from oil fractions (F1–F4) as compared to whole oil was advantageous, as evidenced by the great oxidative stabilities of oil fractions (F1–F4). Cloud point and pour point are important physical properties of any liquid fuel. Some fractions produced under vacuum produced high-quality biodiesel with comparatively much better cloud, pour points and Cold Filter Plugging Point than *Azadirachta indica* oil (Table 6). In fractionation, oil was separated into various fractions. It was expected that some fractions will have better properties than oil and others. As oil comprises all components so it displayed intermediate properties.

Cetane number (CN) is a biofuel quality parameter. CN is one of the factors that determines the ignition capability and depends on the fatty acid or methyl ester composition in the biodiesel. It influences the engine efficiency and ignition power after injection of biofuel [38]. In the present study, the CN value ranged from 9.1–87.3 (Table 7). None of biodiesel sample produced from pure *Azadirachta indica* oil met standard biodiesel cetane number requirements. However, the fractionation process improved the centane number of produced biodiesels. Cetane number of all biodiesel samples produced from *Azadirachta indica* oil fractions was in the recommended range. Biodiesel offers the advantage of a direct replacement for ordinary diesel. In the present study, we propose a much cheaper and simpler method named high vacuum fractional distillation as an alternative to refining to produce high quality biodiesel with desired fuel properties. The cetane number of commonly occurring fatty acids methyl esters are as follows: palmitic (85.9), palmitoleic (51.0), stearic (101.0), oleic (59.3), and linoleic (38.2).

**Table 5.** Iodine values of biodiesel from *Azadirachta indica* seed oil.

| Catalyst | Conc. of Catalyst (%) | Iodine Value | | | | | Oxidative Stability at 110 °C (Induction Time/h) | | | |
|---|---|---|---|---|---|---|---|---|---|---|
| | | *Azadirachta indica* oil | F-1 | F-2 | F-3 | F-4 | *Azadirachta indica* oil | F-1 F-2 | F-3 | F-4 |
| HCl | 20 | 115.2 ± 2.5 | 51.2 ± 1.1 | 66.3 ± 1.3 | 44.1 ± 0.9 | 37.2 ± 0.5 | 11.4 | 28.821.3 | 18.7 | 13.4 |
| | 40 | 129.6 ± 2.1 | 52.4 ± 1.1 | 69.6 ± 1.4 | 44.3 ± 0.9 | 40.2 ± 0.6 | 11.8 | 28.722.4 | 19.4 | 13.2 |
| | 60 | 126.4 ± 2.2 | 60.4 ± 1.3 | 70.0 ± 1.8 | 48.4 ± 0.8 | 41.8 ± 0.4 | 11.7 | 27.923.1 | 19.3 | 13.3 |
| | 80 | 110.4 ± 3.1 | 61.2 ± 1.3 | 71.2 ± 1.9 | 49.2 ± 0.5 | 43.6 ± 0.6 | 11.6 | 26.922.7 | 19.2 | 13.5 |
| | 100 | 118.0 ± 1.2 | 61.4 ± 1.4 | 72.1 ± 2.5 | 49.3 ± 0.4 | 45.2 ± 0.4 | 11.8 | 29.822.5 | 19.1 | 13.9 |
| KOH | 0.2 | 192.0 ± 2.7 | 62.0 ± 1.3 | 63.2 ± 1.9 | 25.2 ± 0.5 | 41.2 ± 0.4 | 11.3 | 28.422.6 | 19.4 | 13.8 |
| | 0.4 | 103.2 ± 1.9 | 66.0 ± 1.6 | 64.0 ± 1.7 | 43.6 ± 0.4 | 46.4 ± 0.6 | 11.4 | 28.322.5 | 18.9 | 13.8 |
| | 0.6 | 184.8 ± 2.6 | 67.3 ± 1.6 | 65.2 ± 1.5 | 55.2 ± 0.4 | 47.5 ± 0.6 | 11.3 | 27.722.8 | 18.7 | 13.7 |
| | 0.8 | 218.4 ± 3.1 | 68.1 ± 1.7 | 71.2 ± 1.5 | 79.2 ± 2.1 | 48.2 ± 0.7 | 11.7 | 27.521.8 | 18.9 | 13.6 |
| | 1 | 201.6 ± 3.5 | 69.6 ± 1.7 | 71.4 ± 1.6 | 91.2 ± 2.6 | 48.8 ± 0.6 | 11.8 | 27.322.6 | 18.7 | 13.8 |
| Lipase | 1 | 110.4 ± 2.5 | 67.2 ± 1.6 | 62.0 ± 1.5 | 55.2 ± 1.8 | 38.0 ± 0.5 | 11.3 | 28.922.8 | 18.6 | 13.5 |
| | 2 | 109.9 ± 2.6 | 68.1 ± 1.6 | 66.3 ± 1.7 | 55.2 ± 1.6 | 38.4 ± 0.7 | 11.8 | 28.122.7 | 18.8 | 13.2 |
| | 3 | 113.9 ± 1.6 | 69.1 ± 1.7 | 68.0 ± 1.8 | 55.2 ± 1.7 | 43.2 ± 0.7 | 11.6 | 28.222.8 | 18.9 | 13.8 |
| | 4 | 114.6 ± 1.7 | 69.2 ± 1.5 | 69.2± 1.8 | 58.4 ± 1.9 | 45.2 ± 0.8 | 11.8 | 28.522.7 | 18.6 | 13.4 |
| | 5 | 129.6 ± 1.9 | 69.8 ± 1.5 | 68.8 ± 1.5 | 59.3 ± 1.8 | 47.2 ± 0.9 | 11.6 | 28.422.6 | 18.8 | 13.6 |

**Table 6.** CP, PP and CFPP (°C) of various biodiesel samples.

| SR. | Catalyst | Conc. % | *Azadirachta indica* Oil | | | F-1 | | | F-2 | | | F-3 | | | F-4 | | |
|---|---|---|---|---|---|---|---|---|---|---|---|---|---|---|---|---|---|
| | | | CP | PP | CFPP | CP | PP | CFPP | CP | PP | CFPP | CP | PP | CFPP | CP | PP | CFPP |
| 1 | | 20 | 2.1 | 0.2 | −3.1 | 0.1 | −4.5 | −10.1 | 1.2 | 0.4 | −6.2 | 6.3 | 1.2 | −5.3 | 2.3 | 0.5 | −3.3 |
| 2 | | 40 | 1.5 | −1.3 | −2.9 | 1.4 | −1.2 | −10.2 | 1.1 | −0.7 | −6.2 | 5.4 | 0.9 | −5.2 | 0.4 | −0.4 | −3.2 |
| 3 | HCl | 60 | 1.1 | −1.8 | −3.0 | 2.1 | −1.9 | −10.0 | −0.3 | -2.5 | −6.3 | 4.3 | 0.8 | −5.0 | 0.8 | −1.2 | −3.0 |
| 4 | | 80 | 1.0 | −1.1 | −2.9 | 2.3 | −3.2 | −10.3 | 3.8 | 0.1 | −6.3 | 4.7 | 0.9 | −5.3 | 0.1 | −0.3 | −3.2 |
| 5 | | 100 | 0.9 | −0.4 | −2.8 | 2.7 | −0.7 | −10.4 | 1.3 | −0.5 | −6.1 | 5.9 | 1.1 | −5.4 | 1.2 | −0.7 | −3.1 |
| 6 | | 0.2 | −0.4 | −4.6 | −5.1 | −0.1 | −4.5 | −11.1 | −2.6 | −4.9 | −7.1 | 1.6 | −1.3 | −5.1 | −0.9 | −4.6 | −4.1 |
| 7 | | 0.4 | −0.6 | −4.8 | −5.0 | −0.3 | −3.4 | −11.0 | 0.1 | −1.1 | −6.0 | 9 | 4.5 | −5.0 | 0.4 | −1.3 | −4.0 |
| 8 | KOH | 0.6 | −0.4 | −4.2 | −5.0 | −2.1 | −4.5 | −11.1 | 0.4 | −1.2 | −6.2 | 8.4 | 4.7 | −5.2 | 0.6 | −2.3 | −4.1 |
| 9 | | 0.8 | −0.5 | −4.7 | −5.8 | −2.5 | −4.8 | −11.2 | 0.1 | −1.1 | −6.2 | 2.7 | −0.3 | −5.2 | 0.7 | −2.5 | −4.2 |
| 10 | | 1 | −0.3 | −4.4 | −5.6 | −2.4 | −4.4 | −11.3 | 0.3 | −1.0 | −6.3 | 10 | 6.6 | −5.2 | 0.8 | −2.7 | −4.3 |
| 11 | | 1 | 1.1 | −3.1 | −3.4 | 7.2 | 3.8 | −12.0 | 5.0 | 3.5 | −6.0 | 15 | 11 | −5.0 | 1.7 | −2.8 | −4.0 |
| 12 | | 2 | 2.1 | −0.1 | −2.9 | 5.2 | 3.5 | −12.2 | 5.2 | 3.0 | −6.2 | 14 | 10 | −5.1 | 2.3 | −0.2 | −4.2 |
| 13 | Lipase | 3 | 2.8 | −3.1 | −3.3 | 8.2 | 3.4 | −12.0 | 5.4 | 2.8 | −6.0 | 15 | 13 | −5.0 | 1.3 | −3.1 | −4.0 |
| 14 | | 4 | 3.8 | 2.9 | −2.9 | 6.2 | 3.5 | −12.9 | 8.4 | 5.2 | −6.9 | 15 | 12 | −4.9 | 1.1 | −0.7 | −4.7 |
| 15 | | 5 | 5.6 | 5.1 | −2.7 | 5.1 | 3.2 | −12.3 | 9.2 | 5.0 | −6.3 | 13 | 11 | −5.1 | 3.5 | 2.3 | −4.6 |
| | | | Diesel oil (CP, °C) | | | Diesel oil (PP, °C) | | | | | | Diesel oil (CFPP, °C) | | | | | |
| | | | −15 to −5 | | | −35 to −15 | | | | | | −8 | | | | | |

**Table 7.** Cetane numbers of biodiesel produced from *Azadirachta indica* seed oil and its various fractions.

| Catalyst | Conc. of Catalyst (%) | Cetane Number 47 Minimum (ASTMD6751) | | | | | Kinematic Viscosity at 40 °C ($mm^2/s$) 3.5–5.0 (ASTMD6751) | | | | | Calorific Value (MJ/kg) | | | | |
|---|---|---|---|---|---|---|---|---|---|---|---|---|---|---|---|---|
| | | *Azadirachta indica oil* | F-1 | F-2 | F-3 | F-4 | *Azadirachta indica oil* | F-1 | F-2 | F-3 | F-4 | *Azadirachta indica oil* | F-1 | F-2 | F-3 | F-4 |
| | 20 | 37.8 | 58.9 | 48.8 | 60.5 | 55.3 | 4.8 | 4.3 | 4.4 | 4.5 | 4.7 | 37.13 | 36.52 | 37.37 | 38.31 | 39.48 |
| | 40 | 34.5 | 57.2 | 65.0 | 61.9 | 68.9 | 4.8 | 4.3 | 4.4 | 4.5 | 4.7 | 37.15 | 36.52 | 37.41 | 38.36 | 39.53 |
| HCl | 60 | 46.8 | 64.4 | 56.2 | 86.1 | 68.6 | 4.8 | 4.3 | 4.3 | 4.5 | 4.6 | 37.15 | 36.59 | 37.41 | 38.38 | 39.56 |
| | 80 | 45.5 | 55.2 | 57.7 | 85.9 | 60.6 | 4.7 | 4.2 | 4.3 | 4.4 | 4.6 | 37.20 | 36.60 | 37.43 | 38.39 | 39.58 |
| | 100 | 36.3 | 73.5 | 55.7 | 69.6 | 83.1 | 4.7 | 4.2 | 4.2 | 4.4 | 4.5 | 37.21 | 36.60 | 37.49 | 38.45 | 39.59 |
| | 0.2 | 17.9 | 51.5 | 49.5 | 57.2 | 59.7 | 4.8 | 4.3 | 4.4 | 4.5 | 4.7 | 37.25 | 36.50 | 37.31 | 38.33 | 39.45 |
| | 0.4 | 36.3 | 55.3 | 48.5 | 53.4 | 54.1 | 4.7 | 4.2 | 4.4 | 4.4 | 4.6 | 37.28 | 36.51 | 37.32 | 38.35 | 39.51 |
| KOH | 0.6 | 17.1 | 63.3 | 48.2 | 51.3 | 72.8 | 4.7 | 4.2 | 4.4 | 4.4 | 4.6 | 37.29 | 36.53 | 37.38 | 38.36 | 39.54 |
| | 0.8 | 9.1 | 57.5 | 55.9 | 45.9 | 52.8 | 4.7 | 4.2 | 4.3 | 4.4 | 4.6 | 37.31 | 36.55 | 37.39 | 38.38 | 39.56 |
| | 1 | 12.6 | 58.4 | 52.9 | 44.0 | 56.7 | 4.7 | 4.2 | 4.3 | 4.3 | 4.4 | 37.34 | 36.56 | 37.40 | 38.41 | 39.56 |
| | 1 | 44.1 | 62.9 | 56.5 | 55.2 | 78.8 | 4.8 | 4.3 | 4.4 | 4.5 | 4.7 | 37.19 | 36.57 | 37.40 | 38.32 | 39.47 |
| | 2 | 46.7 | 58.4 | 72.5 | 65.8 | 67.1 | 4.8 | 4.3 | 4.2 | 4.5 | 4.7 | 37.21 | 36.57 | 37.41 | 38.32 | 39.51 |
| Lipase | 3 | 46.3 | 65.1 | 56.7 | 74.9 | 87.3 | 4.7 | 4.2 | 4.2 | 4.4 | 4.6 | 37.24 | 36.59 | 37.46 | 38.35 | 39.52 |
| | 4 | 46.1 | 68.2 | 60.2 | 70.9 | 86.9 | 4.7 | 4.1 | 4.2 | 4.4 | 4.5 | 37.25 | 36.59 | 37.48 | 38.35 | 39.55 |
| | 5 | 44.6 | 62.3 | 53.5 | 70.7 | 76.7 | 4.7 | 4.1 | 4.2 | 4.3 | 4.5 | 37.26 | 36.60 | 37.48 | 38.39 | 39.56 |
| | | Diesel oil | | | | | Diesel oil | | | | | Diesel oil | | | | |
| | | 46 | | | | | 2.0–4.5 ($mm^2/s$) at 40 °C | | | | | 44.80 MJ/kg | | | | |

A previous study [39] was undertaken with similar objectives such as development of cost-effective processing technology, and elimination of degumming, esterification and dehydration steps from the oil purification processes by conducting a supercritical extraction using $CO_2$ [39].

However, when capital investment and processing cost of high vacuum fractional distillation (HVFD) used in the present study is compared with a supercritical extraction using $CO_2$, HVFD is a much cheaper and simpler alternative. HVFD does not add to the cost of biodiesel as it is an alternative route for oil purification. In another previous study [40], supercritical fluid carbon dioxide was used to separate out biodiesel fraction from impure fatty acid methyl ester (FAME) solution mixes. The disadvantage of doing fractionation after production of methyl esters is the production of a number of useless byproducts, which does not happen in HVFD. Secondly, the prior methodology does not bring any significant improvements in the cetane number, cloud point or pour point of biodiesel.

## 4. Conclusions

The following important conclusions can be drawn from the present study:

- *Azadirachta indica* seed oil can be explored as a renewable and reliable source for biodiesel production.
- High vacuum fractional distillation (HVFD) process was very effective in producing such fractions of *Azadirachta indica* oil which produced better quality biodiesel.
- A superior quality biodiesel was produced from some fractions as presented by oxidative stabilities, iodine values, cloud points, pour points and cetane numbers.
- Biodiesel produced after fractionation of crude oil offers the advantage that it can be used as a direct replacement for ordinary diesel.

**Author Contributions:** Conceptualization, B.I.; M.A.H. and U.R.; methodology, B.I.; M.A.H. M.Z. and U.R.; validation, M.A.H. and I.A.N. and U.R.; formal analysis, M.A.H.; H.N. and M.Z.; investigation, M.A.H. U.R.; Z.M. and M.Z.; resources, M.A.H. and U.R.; data curation, M.A.H. and M.Z.; writing—original draft preparation, B.I.; M.A.H. and U.R.; writing—review and editing, T.S.Y.C.; I.A.N.; U.R. and M.A.H.; visualization, M.A.H. and U.R.; supervision, M.A.H.; H.N. and U.R.; project administration, M.A.H. and U.R.; funding acquisition, U.R. All authors have read and agreed to the published version of the manuscript.

**Funding:** The authors acknowledge their gratitude to King Saud University (Riyadh, Saudi Arabia) for the support of this research through Researchers Supporting Project number (RSP-2019/80).

**Conflicts of Interest:** The authors declare no conflict of interest.

## References

1. Hanif, M.A.; Nisar, S.; Akhtar, M.N.; Nisar, N.; Rashid, N. Optimized production and advanced assessment of biodiesel: A review. *Int. J. Energy Res.* **2018**, *42*, 2070–2083. [CrossRef]
2. Azeem, M.W.; Hanif, M.A.; Al-Sabahi, J.N.; Khan, A.A.; Naz, S.; Ijaz, A. Production of biodiesel from low priced, renewable and abundant date seed oil. *Renew. Energy* **2016**, *86*, 124–132. [CrossRef]
3. Hasan, M.M.; Rahman, M.M. Performance and emission characteristics of biodiesel–diesel blend and environmental and economic impacts of biodiesel production: A review. *Renew. Sustain. Energy Rev.* **2017**, *74*, 938–948. [CrossRef]
4. Knothe, G.; Razon, L.F. Biodiesel fuels. *Prog. Energy Combust. Sci.* **2017**, *58*, 36–59. [CrossRef]
5. Liu, Y.; Tu, Q.; Knothe, G.; Lu, M. Direct transesterification of spent coffee grounds for biodiesel production. *Fuel* **2017**, *199*, 157–161. [CrossRef]
6. Naylor, R.L.; Higgins, M.M. The political economy of biodiesel in an era of low oil prices. *Renew. Sustain. Energy Rev.* **2017**, *77*, 695–705. [CrossRef]
7. Patel, A.; Arora, N.; Mehtani, J.; Pruthi, V.; Pruthi, P.A. Assessment of fuel properties on the basis of fatty acid profiles of oleaginous yeast for potential biodiesel production. *Renew. Sustain. Energy Rev.* **2017**, *77*, 604–616. [CrossRef]
8. Rehan, M.; Gardy, J.; Demirbas, A.; Rashid, U.; Budzianowski, W.M.; Pant, D.; Nizami, A.S. Waste to biodiesel: A preliminary assessment for Saudi Arabia. *Bioresource Technol.* **2018**, *250*, 17–25.

9. Bhatti, H.N.; Hanif, M.A.; Faruq, U.; Sheikh, M.A. Acid and base catalyzed transesterification of animal fats to biodiesel. *Iran. J. Chem. Chem. Eng.* **2008**, *27*, 41–48.
10. Bhatti, H.N.; Hanif, M.A.; Qasim, M. Biodiesel production from waste tallow. *Fuel* **2008**, *87*, 2961–2966. [CrossRef]
11. Budžaki, S.; Miljić, G.; Tišma, M.; Sundaram, S.; Hessel, V. Is there a future for enzymatic biodiesel industrial production in microreactors? *Appl. Energy* **2017**, *201*, 124–134. [CrossRef]
12. Ikram, M.M.; Hanif, M.A.; Khan, G.S.; Rashid, U.; Nadeem, F. Significant Seed Oil Feedstocks for Renewable Production of Biodiesel: A Review. *Curr. Org. Chem.* **2019**, *23*, 1509–1516. [CrossRef]
13. Inam, S.; Khan, S.; Nadeem, F. Impacts of Derivatization on Physiochemical Fuel Quality Parameters of Fatty Acid Methyl Esters (FAME)-A Comprehensive Review. *Int. J. Chem. Biochem. Sci.* **2019**, *15*, 42–49.
14. Kalsoom, M.; El Zerey-Belaskri, A.; Nadeem, F.; Shehzad, M.R. Fatty Acid Chain Length Optimization for Biodiesel Production Using Different Chemical and Biochemical Approaches–A Comprehensive. *Int. J. Chem. Biochem. Sci.* **2017**, *11*, 75–94.
15. Nadeem, F.; Inam, S.; Rashid, U.; Kainat, R.; Iftikhar, A. A Review of Geographical Distribution, Phytochemistry, Biological Properties and Potential Applications of Pongamia pinatta. *Int. J. Chem. Biochem. Sci.* **2016**, *10*, 79–86.
16. Nadeem, F.; Shahzadi, A.; El Zerey-Belaskri, A.; Abbas, Z. Conventional and Advanced Purification Techniques for Crude Biodiesel–A Critical Review. *Int. J. Chem. Biochem. Sci.* **2017**, *12*, 113–121.
17. Shahzadi, A.; Grondahl, L.; Nadeem, F. Development of Effective Composite Supports for Production of Biodiesel-A Detailed Review. *Int. J. Chem. Biochem. Sci.* **2019**, *16*, 76–86.
18. Waseem, H.H.; El Zerey-Belaskri, A.; Nadeem, F.; Yaqoob, I. The Downside of Biodiesel Fuel–A Review. *Int. J. Chem. Biochem. Sci.* **2016**, *9*, 97–106.
19. Yaqoob, I.; Rashid, U.; Nadeem, F. Alumina Supported Catalytic Materials for Biodiesel Production-A Detailed Review. *Int. J. Chem. Biochem. Sci.* **2019**, *16*, 41–53.
20. Hanif, M.A.; Nisar, S.; Rashid, U. Supported solid and heteropoly acid catalysts for production of biodiesel. *Catal. Rev.* **2017**, *59*, 165–188. [CrossRef]
21. Mehboob, A.; Nisar, S.; Rashid, U.; Choong, T.S.Y.; Khalid, T.; Qadeer, H.A. Reactor designs for the production of biodiesel. *Int. J. Chem. Biochem. Sci.* **2016**, *10*, 87–94.
22. Nisar, S.; Ishaq, A.; Asia Sultana, F.; Shehzad, M.R. Reactions other than transesterification for biodiesel production. *Int. J. Chem. Biochem. Sci.* **2017**, *12*, 141–146.
23. Xue, Y.; Zhao, W.; Ma, P.; Zhao, Z.; Xu, G.; Yang, C.; Chen, H.; Lin, H.; Han, S. Ternary blends of biodiesel with petro-diesel and diesel from direct coal liquefaction for improving the cold flow properties of waste cooking oil biodiesel. *Fuel* **2016**, *177*, 46–52. [CrossRef]
24. Ali, M.H.; Mashud, M.; Rubel, M.R.; Ahmad, R.H. Biodiesel from Neem oil as an alternative fuel for Diesel engine. *Procedia Eng.* **2013**, *56*, 625–630. [CrossRef]
25. Junior, I.I.; Flores, M.C.; Sutili, F.K.; Leite, S.G.F.; de Miranda, L.S.; Leal, I.C.R.; de Souza, R.O.M.A. Fatty acids residue from palm oil refining process as feedstock for lipase catalyzed monoacylglicerol production under batch and continuous flow conditions. *J. Mol. Catal. B* **2012**, *77*, 53–58. [CrossRef]
26. Datta, A.; Mandal, B.K. A comprehensive review of biodiesel as an alternative fuel for compression ignition engine. *Renew. Sustain. Energy Rev.* **2016**, *57*, 799–821. [CrossRef]
27. Muthu, H.; SathyaSelvabala, V.; Varathachary, T.; Kirupha Selvaraj, D.; Nandagopal, J.; Subramanian, S. Synthesis of biodiesel from Neem oil using sulfated zirconia via tranesterification. *Braz. J. Chem. Eng.* **2010**, *27*, 601–608. [CrossRef]
28. De Normalisation, C.E. Automotive Fuels–Diesel–Requirements and Test Methods. *SS-EN* **2014**, *590*, 2013.
29. Meira, M.; Quintella, C.M.; dos Santos Tanajura, A.; Da Silva, H.R.G.; Fernando, J.D.E.S.; da Costa Neto, P.R.; Pepe, I.M.; Santos, M.A.; Nascimento, L.L. Determination of the oxidation stability of biodiesel and oils by spectrofluorimetry and multivariate calibration. *Talanta* **2011**, *85*, 430–434. [CrossRef]
30. Al-Farsi, M.A.; Lee, C.Y. Optimization of phenolics and dietary fibre extraction from date seeds. *Food Chem.* **2008**, *108*, 977–985. [CrossRef]
31. Rashid, U.; Anwar, F. Production of biodiesel through optimized alkaline-catalyzed transesterification of rapeseed oil. *Fuel* **2008**, *87*, 265–273. [CrossRef]
32. Nisar, S.; Farooq, U.; Idress, F.; Rashid, U.; Choong, T.S.Y.; Umar, A. Biodiesel by-product glycerol and its products. *Int. J. Chem. Biochem. Sci.* **2016**, *9*, 92–96.

33. Sagiroglu, A.; Isbilir, S.Ş.; Ozcan, M.H.; Paluzar, H.; Toprakkiran, N.M. Comparison of biodiesel productivities of different vegetable oils by acidic catalysis. *Chem. Ind. Chem. Eng. Q.* **2011**, *17*, 53–58. [CrossRef]
34. Canakci, M.; Van Gerpen, J. Biodiesel production from oils and fats with high free fatty acids. *Trans. ASAE* **2001**, *44*, 1429. [CrossRef]
35. Velmurugan, K.; Sathiyagnanam, A.P. Effect of biodiesel fuel properties and formation of NOx emissions: A review. *Int. J. Ambient Energy* **2017**, *38*, 644–651. [CrossRef]
36. Azam, M.M.; Waris, A.; Nahar, N. Prospects and potential of fatty acid methyl esters of some non-traditional seed oils for use as biodiesel in India. *Biomass Bioenergy* **2005**, *29*, 293–302.
37. Canakci, M.; Sanli, H. Biodiesel production from various feedstocks and their effects on the fuel properties. *J. Ind. Microbiol. Biotechnol.* **2008**, *35*, 431–441. [CrossRef]
38. Bamgboye, A.; Hansen, A. Prediction of cetane number of biodiesel fuel from the fatty acid methyl ester (FAME) composition. *Int. Agrophys.* **2008**, *22*, 21.
39. Fernández, C.M.; Fiori, L.; Ramos, M.J.; Pérez, Á.; Rodríguez, J.F. Supercritical extraction and fractionation of Jatropha curcas L. oil for biodiesel production. *J. Supercrit. Fluids* **2015**, *97*, 100–106. [CrossRef]
40. Wei, C.-Y.; Huang, T.-C.; Yu, Z.-R.; Wang, B.-J.; Chen, H.-H. Fractionation for biodiesel purification using supercritical carbon dioxide. *Energies* **2014**, *7*, 824–833. [CrossRef]

Article

# Outlook for Direct Use of Sunflower and Castor Oils as Biofuels in Compression Ignition Diesel Engines, Being Part of Diesel/Ethyl Acetate/Straight Vegetable Oil Triple Blends

**Laura Aguado-Deblas [1], Rafael Estevez [1,*], Jesús Hidalgo-Carrillo [1], Felipa M. Bautista [1], Carlos Luna [1], Juan Calero [1], Alejandro Posadillo [2], Antonio A. Romero [1] and Diego Luna [1]**

[1] Departamento de Química Orgánica, Universidad de Córdoba, Campus de Rabanales, Ed. Marie Curie, 14014 Córdoba, Spain; aguadolaura8@gmail.com (L.A.-D.); Jesus.hidalgo@uco.es (J.H.-C.); qo1baruf@uco.es (F.M.B.); qo2luduc@uco.es (C.L.); p72camaj@gmail.com (J.C.); qo1rorea@uco.es (A.A.R.); diego.luna@uco.es (D.L.)

[2] Seneca Green Catalyst S.L., Campus de Rabanales, 14014 Córdoba, Spain; seneca@uco.es

* Correspondence: q72estor@uco.es; Tel.: +34-957212065

Received: 9 August 2020; Accepted: 7 September 2020; Published: 16 September 2020

**Abstract:** Today, biofuels are indispensable in the implementation of fossil fuels replacement processes. This study evaluates ethyl acetate (EA) as a solvent of two straight vegetable oils (SVOs), castor oil (CO), and sunflower oil (SO), in order to obtain EA/SVO double blends that can be used directly as biofuels, or along with fossil diesel (D), in the current compression-ignition (C.I.) engines. The interest of EA as oxygenated additive lies not only in its low price and renewable character, but also in its very attractive properties such as low kinematic viscosity, reasonable energy density, high oxygen content, and rich cold flow properties. Revelant fuel properties of EA/SVO double and D/EA/SVO triple blends have been object of study including kinematic viscosity, pour point (PP), cloud point (CP), calorific value (CV), and cetane number (CN). The suitability of using these blends as fuels has been tested by running them on a diesel engine electric generator, analyzing their effect on engine power output, fuel consumption, and smoke emissions. Results obtained indicate that the D/EA/SO and D/EA/CO triple blends, composed by up to 24% and 36% EA, respectively, allow a fossil diesel substitution up to 60–80% providing power values very similar to conventional diesel.In addition, in exchange of a slight fuel consumption, a very notable lessening in the emission of pollutants as well as a better behavior at low temperatures, as compared to diesel, are achieved.

**Keywords:** ethyl acetate; castor oil; sunflower oil; straight vegetable oils; vegetable oil blends; biofuels; diesel engine; soot emissions; engine power output

---

## 1. Introduction

Nowadays, fossil fuels such as coal, oil, and natural gas represent approximately 80% of the world energy consumption. Furthermore, it is expected that the industrialization, together with the growing population will increase energy demand in the next years, with the consequent negative effects on global climate [1]. Thus, it is mandatory the use of renewable energy sources, such as solar, wind, and biofuels, to avoid damaging the environment [2]. In this scenario, the production of fuels from biomass has become as one of the best options to provide renewable liquid fuels for transportation. On one hand, until today, biodiesel has been the mostly employed biofuel as substitute of fossil diesel. Biodiesel is produced through transesterification of triglycerides from vegetable oils (palm, corn, soybean, sunflower, etc.) with methanol or ethanol, being a sustainable and environmentally friendly alternative. Furthermore, it has excellent properties operating in the current diesel engines, not only by itself but also in blends with fossil diesel. The main drawback in the biodiesel production is the generation of glycerol as by-product in around 10 wt. % of the total biodiesel, which makes the biodiesel production on an industrial scale economically unfeasible. Hence, the competitive commercialization of biodiesel is still a challenge [3].

On the other hand, the use of SVOs as fuel was already initiated by Rudolph Diesel, when he made the first demonstration of his compression ignition engine at the 1900 Paris Exposition with peanut oil, and their usage continued until the 1920s, when diesel engine was adjusted for running with a fraction of fossil petroleum (currently known as No. 2 diesel fuel) [4]. Nowadays, SVOs are considered a very attractive alternative to fossil fuels because of their renewable nature and their high availability. Moreover, some vegetable oils exhibit physicochemical properties analogous to conventional diesel, excluding their high kinematic viscosity that lead to engine problems (poor fuel atomization and coke formation). That is the reason why the use of vegetable oils as drop-in biofuels requires previous treatments (i.e., transesterification reaction to obtain biodiesel) that reduce their viscosity values.

Recently, it has emerged a new methodology, to the direct usage of SVOs as biofuels in C.I. engines, consisting of blending vegetable oil with a lower viscosity solvent (LVS), instead of undergoing them chemical process. This approach has the advantage of avoiding the energetic and economic costs associated to the transesterification process for the biodiesel production. Thereby, the high viscosity of oils can be reduced to limits imposed by EN 590 standard to operate in the present diesel engines. In this respect, an intensive investigation about different LVSs used as fuels has been reported in literature, including different alcohols (methanol, ethanol, and n-butanol) [5–8] and light vegetable oils (eucalyptus, camphor, orange, and pine oils) [9–12]. Generally, these compounds have been designated as LVLC (low viscous low cetane), since they exhibit low viscosity and energy density values and sometimes, low cetane number, as compared to conventional diesel [13].

Following this strategy, the low viscosity of gasoline has been harnessed to use it as LVS, in mixtures with SVOs from several seeds (canola, sunflower, camelina, and carinata). In this way, SVO/gasoline blends, with a content of gasoline ranging between 10 and 30% in volume, have allowed a higher level of fossil fuel substitution than their counterparts blends using fossil diesel. Aside from adequate density and viscosity values for being successfully employed in a diesel engine, the thermal efficiency of these blends was close to conventional diesel, although the biofuel consumption was approximately 10% higher than that of diesel fossil [14]. Gasoline/SVO blending methodology has been more recently applied even with castor oil, which has a high viscosity value of 226.2 cSt. Furthermore, this is non-edible oil, avoiding ethical problems [15]. Thus, it has been reported that diesel/gasoline/castor oil triple blends generate similar perform to fossil diesel, allowing a substitution of fossil fuel up to 24%. For its part, the same triple blend using sunflower oil provided 36% of fossil fuel substitution. Other advantages obtained with these diesel/gasoline/SVO blends were a considerably reduction in smoke opacity as well as a significative improvement of flow properties at cold climates [15].

In addition to gasoline, diethyl ether (DEE) [16] and acetone (ACE) [17] have been employed as effective LVS of CO and SO, in triple blends with diesel. Both additives can be obtained from ethanol which, in turn, can be obtained from biomass, making the process more sustainable. Both of these oxygenated additives were able to lower soot emissions considerably, maintaining a good engine performance. In spite of favorable fuel properties of DEE and ACE [16,17], the low calorific power was revealed as the limiting factor to incorporate percentages of biofuel LVS/SVO higher than 40%. Several studies have also been published fueling a diesel engine with diesel/DEE/SVOs triple blends. Among the oils employed, it can be highlighted cashew nut shell oil [18], bael oil [19], as well as sunflower and castor oil [16].

Additionally, ethyl acetate has been evaluated as a potential oxygenated biofuel since can be readily produced from renewable feedstocks by several processes such as the direct esterification of ethanol with acetic acid [20], through dehydrogenative dimerization of bioethanol with liberation of molecular hydrogen [21], or by a biochemical process that includes acidogenic fermentation of agricultural wastes materials [22]. EA is an environmentally friendly compound that exhibit very low kinematic viscosity, high miscibility with VOs and fossil diesel, high oxygen content, high auto-ignition temperature which makes fuel safer for handling and transportation, and good cold flow properties (very low PP and CP values) to improve winter engine performance. Hence, the application of EA as solvent of second-generation oils, i.e., castor oil or waste cooking oils, could be an inexpensive and valuable alternative to accomplish the replacement of fossil fuel, which is mandatory to realize now with the operating car fleet.

Prominently improvements have been reported in brake thermal efficiency (BTE) but with a slightly higher brake specific fuel consumption (BSFC) using ethyl acetate as fuel or additive for C.I. engines [23]. In addition, exhaust temperature as well as CO, HC, and NOx emissions have been diminished with this fuel [22–24]. Similarly, the addition of 10% EA to diesel has showed similar brake specific fuel consumption respect to diesel, but with a notable reduction of soot emissions [25]. On the other hand, the addition of ethyl acetate to some bio-oils has provided important advantages to enhance the storage stability, thanks to its non-hygroscopic nature [26]. This compound has been also applied in blends with gasoline, leading to decrease in CO and HC emissions by around 50% due to high oxygen content of ethyl acetate [27]. Moreover, EA has been also used as a surfactant to emulsify a 50% diesel/50% ethanol mixture, resulting in an increase in BTE and a decrease in BSFC, smoke density, particulate matter, exhaust gas temperature, HCs and CO [28].

In this research, the possibility of replacing fossil diesel by renewable biofuels obtained by blending EA with SVOs, either sunflower or castor oil, has been evaluated. First, the optimum EA/SVO blends that meet with viscosity requirements according to European normative, were chosen to be subsequently mixed with diesel fossil. Then, the different EA/SVO double and D/EA/SVO triple blends have been tested in a C.I. engine. Several parameters were measured, i.e., power output, fuel consumption and soot emissions. Moreover, some of the most crucial fuel properties, e.g., calorific value, cetane number, cloud point, pour point, and kinematic viscosity were determined to ascertain the suitability of the mixtures as (bio)fuels. In addition, all these measurements were performed with fossil diesel for comparison purposes.

## 2. Materials and Methods

Table 1 shows the main physicochemical properties of diesel, sunflower oil, castor oil, and ethyl acetate.

**Table 1.** Physicochemical properties of fossil diesel, sunflower oil, castor oil, and ethyl acetate [22,23,27,29,30]. Kinematic viscosity values have been experimentally measured in the present work.

| Property | Diesel | Sunflower Oil | Castor Oil | Ethyl Acetate |
|---|---|---|---|---|
| Density at 15 °C (kg/m$^3$) | 830.0 | 920.0 | 962.0 | 906.0 |
| Kinematic viscosity at 40 °C (cSt) | 3.20 | 37.80 | 226.20 | 0.45 |
| Oxygen content (wt. %) | 0 | 10 | 15 | 36 |
| Calorific value (MJ/L) | 35.52 | 36.43 | 38.00 | 21.55 |
| Flash point (°C) | 66.0 | 220.0 | 228.0 | −4.0 |
| Auto-ignition temperature (°C) | 250 | 316 | 448 | 460 |
| Cetane number | 51 | 37 | 40 | 10 |

### 2.1. *Ethyl Acetate/Vegetable Oil Double Blends and Diesel/Ethyl Acetate/Vegetable Oil Triple Blends*

Sunflower oil was supplied by a local market, castor oil was purchased from Panreac Company (Castellar Del Vallès, Spain), and ethyl acetate was acquired from Sigma-Aldrich Chemical Company (St. Louis, MO, USA, ≥99.8% purity). Fossil diesel, used as reference fuel, was purchased from local Repsol service station. The experimental methodology is shown in Figure 1. The EA/SO and EA/CO double blends, and D/EA/SO and D/EA/CO triple blends were prepared by manual mixing of the different components at room temperature. The mixtures were designated as B20, B40, B60, B80, and B100, where 100% fossil diesel is indicated as B0 and EA/SVO blend is B100, Table 2.

**Figure 1.** Scheme of experimental methodology.

**Table 2.** Diesel/ethyl acetate/sunflower oil blends containing 40% ethyl acetate, and diesel/ethyl acetate/castor oil blends containing 45% ethyl acetate.

| Blend | Nomenclature | B0 | B20 | B40 | B60 | B80 | B100 |
|---|---|---|---|---|---|---|---|
| | % D | 100 | 80 | 60 | 40 | 20 | 0 |
| D/EA/SO | % EA | 0 | 8 | 16 | 24 | 32 | 40 |
| | % SO | 0 | 12 | 24 | 36 | 48 | 60 |
| D/EA/CO | % EA | 0 | 9 | 18 | 27 | 36 | 45 |
| | % CO | 0 | 11 | 22 | 33 | 44 | 55 |

### 2.2. Characterization of the (Bio)fuel Blends

A fuel is defined by a series of properties that determine its performance in the engine, and consequently, its commercialization. Blends properties, including kinematic viscosity, pour point, cloud point, calorific value and cetane number, were determined through predictive equationsor experimental methodology described below.

#### 2.2.1. Kinematic Viscosity

The existence of maximum and minimum limits for the viscosity of a fuel is required for ensuring the engine works without any risk, and fuel injection system is unaffected [31]. The viscosity measures of mixtures were carried out in an Ostwald-Cannon-Fenske capillary viscometer (Proton Routine Viscometer 33,200, size 150) at 40 °C, according to the European standard EN ISO 3104 test method and following the methodology described in previous works [16,17].

#### 2.2.2. Pour Point (PP) and Cloud Point (CP)

Cloud point and pour point are cold flow properties responsible for solidification of fuel at cold operating conditions. At low temperatures, fuel crystallization results in clog to fuel lines and filters, which causes problems to start engine as a consequence of the lack of fuel [32].

CP and PP measurements have been performed following the EN 23015/ASTM D-2500 and ISO 3016/ASTM D97 standards, respectively. The experiments are performed in accordance with the procedure described in previous investigations [16,17]. All results are presented as average values of duplicate determinations and error is calculated as standard deviation.

#### 2.2.3. Calorific Value

Calorific value (CV) is a crucial fuel property since it is directly related to power output of the engine [30]. This parameter has been determined following the Kay Mixing rule (Equation (1)):

$$CV = \sum_i CViXi \quad (1)$$

where $CVi$ is the calorific value of each component and $Xi$ is the volumetric fraction of every component [33].

#### 2.2.4. Cetane Number

Cetane number (CN) is one of key parameters to define the ignition quality of a fuel in a diesel engine. In general, CN of fuels must be above 51, as according norm EN 590, to facilitate autoignition and provide short ignition delay. However, CN values too high can cause the ignition delay is very short and combustion may start before the fuel and air are properly mixed, leading to an incomplete combustion [34]. Experimental determination of the CN of a fuel is a procedure tedious and expensive, and therefore estimation of cetane number of mixtures is carried out using the following simple equation:

$$CN = \sum_i CNiXi \quad (2)$$

where $CNi$ is the cetane number of each component and $Xi$ is the volumetric fraction of every component [33].

### *2.3. Performance of a Diesel Engine-Electrogenerator Set Fuelled with EA/SVO and D/EA/SVO Blends*

The performance and soot emissions of a diesel engine-electric generator set have been analyzed following the same experimental methodology previously reported [15–17]. Engine specifications are shown in Table 3. In addition, the operating conditions of the engine were not modified during the tests.

**Table 3.** Specifications of the Diesel Engine-Electrogenerator Set.

| Model | AYERBE 4000 Diesel |
|---|---|
| Alternator | LINZ-SP 10MF 4.2 KVA |
| Engine | YANMAR LN-70 296 cc 6.7 HP |
| Speed | 3000 rpm |
| Maximum power | 5 KVA |
| Voltage | 230 V |
| Consumption | 1.3 L/H (75%) |

The electrical power ($P$) in watts can be easily calculated using a voltmeter-ammeter by application of Equation (3):

$$P = V \cdot I \quad (3)$$

where $V$ is the potential difference or voltage (in volts) and $I$ is the electric current intensity or amperage (in amps).

The contamination degree (measured in Bosch number units) was determined from the opacity of the smoke generated in the combustion process, using a smoke density tester. In this research, the smoke density was measured by an opacimeter-type TESTO 338 density gauge (or Bosch smoke meter), Figure 2, according to standard method ASTM D-2156. The measurement range for smoke density is 0 to 2.5 with ±0.03 accuracy, where the value 0 represents total clarity on the paper and 2.5 is the corresponding value to 100% cloudy.

**Figure 2.** Mechanical and environmental characterization of diesel engine-electrogenerator set [17].

The fuel consumption of diesel engine (in liters per hour) is calculated supplying to engine an identical fuel volume (0.5 L) of each proposed (bio)fuel and measuring the volume consumed after a specified time. These measurements were performed at three engine loads that are representative of low (1 kW), medium (3 kW), and high (5 kW) power demands of the engine. Tests were done in triplicate and the results are shown as average along with standard deviation, represented as error bars.

## 3. Results and Discussion

### 3.1. Fuel Properties of EA/SVO Double Blends, and D/EA/SVO Triple Blends

The kinematic viscosity results for ethyl acetate/sunflower oil (EA/SO) and ethyl acetate/castor oil (EA/CO) double blends are shown in Table 4. As can be noticed, a remarkable decrease in the viscosity values of SVOs was achieved by the use of ethyl acetate as solvent, as expected. In fact, only a 20% of EA reduce the viscosity of CO from 226.2 cSt to 26.26 cSt, while the same proportion of EA get to decrease the SO viscosity value from 37.80 to 11.52 cSt. Thus, increasing the volume of ethyl acetate in the blends we are able to fulfill the viscosity values for being employed as fuels in C.I. engines, as establish the European standard EN 590 ISO 3104. It is noticeable the stronger influence of EA on castor oil, in comparison with SO, to reduce the very higher viscosity value of this oil. Following the European normative (viscosities between 2.0 and 4.5 cSt), the blends with suitable viscosity values which were selected for the reformulation of diesel are obtained with a proportion 40/60 of EA/SO and 45/55 of EA/CO, Table 4.

**Table 4.** Viscosity values at 40 °C (cSt, centistokes) of ethyl acetate (EA)/sunflower oil (SO) and EA/castor oil (CO) double blends. Errors expressed as standard deviation have been calculated from average three measurement.

| Property | Blend | Ethyl Acetate (% by Volume) | | | | | |
|---|---|---|---|---|---|---|---|
| | | 0 | 20 | 40 | 45 | 50 | 100 |
| Kinematic viscosity (cSt) | EA/SO | 37.80 ± 0.46 | 11.52 ± 0.30 | 4.42 ± 0.04 | 3.76 ± 0.06 | 3.16 ± 0.06 | 0.45 ± 0.01 |
| | EA/CO | 226.20 ± 0.55 | 26.26 ± 0.06 | 5.97 ± 0.07 | 4.47 ± 0.06 | 3.46 ± 0.12 | 0.45 ± 0.01 |

The best EA/SO and EA/CO double blends, containing 40% an 45% of ethyl acetate, respectively, were employed to prepare the D/EA/SVO triple mixtures by addition of different volumetric proportions to fossil diesel, which vary from 20% to 80% (B20 to B80). The kinematic viscosity results of triple mixtures are shown in Figure 3. As can be seen, the incorporation of the EA/SVO biofuels, from B0 to B100, promotes a slight increment in viscosity values of all blends, which was expected since they exhibit higher viscosities than fossil diesel. In general, the viscosities of the blends fulfill with European regulations EN 590, being in the range of 3.23–4.47 cSt.

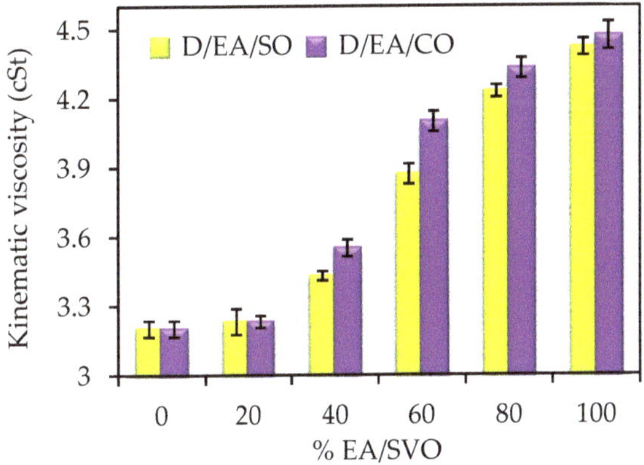


**Figure 3.** Kinematic viscosity values at 40 °C of diesel/ethyl acetate/sunflower oil (D/EA/SO) and diesel/ethyl acetate/castor oil (D/EA/CO) triple blends. The results are the average of three measurements and errors are represented as standard deviation through error bars.

Cloud point, pour point, calorific values, and cetane number of triple blends with SO are collected in Table 5, whereas those with CO are collected in Table 6. Flow properties of diesel at low temperatures are greatly improved by the presence of EA in the D/EA/SVO blends, independently on the vegetable oil employed. In fact, a small percentage of ethyl acetate (8% in B20 blends with SO) generates a significant decrease in the pour point (PP), from −16.0 to −19.9 °C, and in the cloud point (CP), from −6.0 to −12.0 °C. Likewise, the triple blend with 11% of castor oil and 9% of EA reaches a reduction in PP and CP of up to 5 °C and 7 °C, respectively. Based on these results, the blends which exhibit the best behavior at low temperature are composed by 40% of biofuel, i.e., 60/16/24 (D/EA/SO) and 60/18/22 (D/EA/CO), which has conducted to a reduction of up to 4.2–6.0 °C in PP and 7.6–9.8 °C in CP. Particularly, blends with castor oil as SVO allow to obtain a slight improvement of CP and PP than their counterparts with sunflower oil. This behavior is very similar to that found using others compounds such as diethyl ether [16] or acetone [17] as additives in triple mixtures with the SVOs.

**Table 5.** Cloud point, pour point, calorific value and cetane number of diesel/ethyl acetate/sunflower oil triple blends (EA = 40%). Errors have been calculated from average three measurement and expressed as standard deviation.

| Diesel/Ethyl Acetate/Sunflower Oil Blend | | | | | | |
|---|---|---|---|---|---|---|
| Nomenclature (% Renewable) | B0 100/0/0 | B20 80/8/12 | B40 60/16/24 | B60 40/24/36 | B80 20/32/48 | B100 0/40/60 |
| Cloud point (°C) | −6.0 ± 1.0 | −12.0 ± 1.0 | −13.6 ± 0.9 | −12.5 ± 0.5 | −12.2 ± 0.8 | −10.6 ± 1.0 |
| Pour point (°C) | −16.0 ± 1.2 | −19.9 ± 0.8 | −20.2 ± 1.1 | −18.5 ± 0.7 | −18.1 ± 1.0 | −19.0 ± 0.8 |
| Calorific value (MJ/L) | 35.52 | 34.51 | 33.50 | 32.49 | 31.49 | 34.38 |
| Cetane number | 51.00 | 46.04 | 41.08 | 36.12 | 31.16 | 26.20 |

**Table 6.** Cloud point, pour point, calorific value and cetane number of diesel/ethyl acetate/castor oil triple blends (EA = 45%). Errors have been calculated from average three measurements and expressed as standard deviation.

| | Diesel/Ethyl Acetate/Castor Oil Blend | | | | | |
|---|---|---|---|---|---|---|
| Nomenclature (% Renewable) | B0 100/0/0 | B20 80/9/11 | B40 60/18/22 | B60 40/27/33 | B80 20/36/44 | B100 0/45/55 |
| Cloud point (°C) | −6.0 ± 1.0 | −13.0 ± 1.0 | −15.8 ± 0.7 | −12.0 ± 0.6 | −11.0 ± 1.0 | −12.2 ± 1.0 |
| Pour point (°C) | −16.0 ± 1.2 | −21.0 ± 1.0 | −22.0 ± 1.0 | −19.4 ± 0.8 | −18.5 ± 0.6 | −20.0 ± 0.9 |
| Calorific value (MJ/L) | 35.52 | 34.54 | 33.55 | 32.57 | 31.58 | 30.60 |
| Cetane number | 51.00 | 46.10 | 41.20 | 36.30 | 31.40 | 26.50 |

Regarding the calorific values for triple blends (Tables 5 and 6), as the percentage of ethyl acetate in the blend increase, the calorific value decrease. This is logical since EA exhibits a lower calorific value than both, diesel, and SVOs. Therefore, the most favourable results are found in the B20 triple blends, which have the highest calorific values, 34.51 MJ/L for the blends with sunflower oil (2.84% lower than diesel), and 34.54 MJ/L for the blends with castor oil (2.76% lower than diesel), while the B80 triple blends with highest biofuel content (80%) show the lowest calorific values of 31.49 and 31.58 MJ/L, i.e., 11.35 and 11.09% lower than diesel, for SO and CO blends respectively. As can be observed, there is no appreciable differences between calorific values of blends containing either sunflower or castor oil, since both oils display a comparable calorific power (see Table 1). Overall, the calorific value of biofuels EA/SVO studied was around 14% lower than that of diesel (B0).

The results related to cetane number (Tables 5 and 6) of blends with SO and CO were very similar among them, showing a decrease in cetane number values as EA/SVO ratio increases. It can be seen that the cetane number of all blends is below 51, which is the minimum cetane number of diesel in European standard EN 14214. Therefore, the ignition delay of these fuels is expected to be longer respect to conventional diesel fuel.

### 3.2. *Performance of a Diesel Engine Operating as Electric Generator*

According to the characterization results, the mixtures proposed as (bio)fuels that comply with kinematic viscosity requirements stablished by European normative EN 590, were tested in a diesel engine. The engine loads employed for the tests were 0, 1000, 2000, 3000, 4000, and 5000 W. Figure 4 illustrates the impact of engine load on engine performance fueled with the different D/EA/SVO triple blends, containing SO (Figure 4a) and CO (Figure 4b). Furthermore, the performance of EA/SVO (B100) and fossil diesel were also included for comparison.

Generally, as the power supplied increases up to 4000 W, the power output also increases, whereas the highest engine load (5000 W) generated a drop in power output of engine. This behaviour is observed with D/EA/SO blends containing up to 60% of biofuel, and with D/EA/SO blends containing up to 80% of biofuel. On the contrary, blends B80 with SO and EA/SVO double blends display a different trend: firstly, the power generated increases from 0 to 3000 W; then, it falls down from 3000 to 4000 W; and finally, it remains stable when the highest load is applied to engine. It is noteworthy that, at the highest load conditions, i.e., 4000 and 5000 W, the blends D/EA/CO exhibit very similar or even higher power output values than diesel. Thus, fuels composed by up to 60% of biofuel, EA/SO (B20–B60), and up to 80% of biofuel EA/CO (B20–B80) revealed a very notable efficiency on engine. However, higher concentrations of EA/SVO led to a worst engine performance, in comparison with conventional diesel or analogous mixtures with lower biofuel content. Be that as it may, it is very interesting the fact that the EA/SVO double blends allow the running on engine without employing any fossil diesel content, which means a 100% of diesel substitution for renewable compounds. This achievement is even more remarkable for biofuel containing castor oil, since the decrease in the power output is between 7% and 30% in respect to diesel, which is lower than that obtained for biofuel with sunflower oil (50–74% lower than diesel). Usually, triple mixtures with castor oil exhibit better behavior as (bio)fuels than those containing sunflower oil in any proportion investigated. In this sense, the usage of castor oil as part of these triple mixtures not only improves the power results as compared to sunflower oil, but also achieves better results with regard to fossil diesel, even with higher percentages of ethyl acetate up to 36% (B20–B80 blends). Taking into account that both SVOs exhibit very similar calorific values (Table 1), the better behavior of CO could be due to the higher cetane number of this oil, which improves the combustion quality. The progressive reduction in the engine performance observed with the blends B20 to B100 could be attributed to the low calorific value that ethyl acetate exhibit. Hence, a higher proportion of EA in the blend would promote the reduction in the energy content of blends (Table 1). This fact agreed with the results reported in previous investigations where triple blends containing ethanol [6], diethyl ether [16], or acetone [17] were employed as LVLCs. Nonetheless, the influence of other operating parameters on engine performance cannot be ruled out.

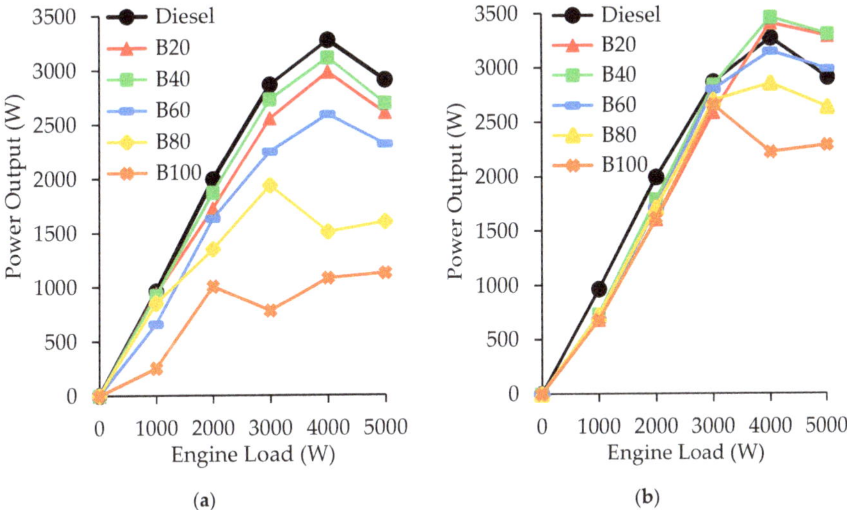


**Figure 4.** Power output (W) at different engine loads (1000–5000 W) for (**a**) diesel/ethyl acetate/sunflower oil; (**b**) diesel/ethyl acetate/castor oil blends. The error in measurements was always less than 3%.

### 3.3. Smoke Emissions: Opacity

Figure 5 shows the smoke emissions (expressed in Bosch number) of the different blends tested. As can be seen, the higher the amount of EA/SVO in the blend, the higher the reduction of soot emissions. As we previously reported, this behavior is explained by the increase of the oxygen content in the fuel. Oxygen acts reducing the formation of rich zones and promoting the oxidation of soot nuclei during fuel combustion. Although the oxygen content is the dominant factor on soot emissions, in turn a lower cetane number has also demonstrated to decrease soot particles emissions due to that a longer ignition delay provides more time for the premixing of fuel and air prior to the start of combustion, which increases the oxidation of soot particles [35].

Accordingly, the lowest opacity values are obtained with pure biofuels EA/SVO (B100), independently on the vegetable oil used, since they have the highest concentration of EA and the lowest CN. These B100 blends decrease soot emissions up to 85% when sunflower oil is employed and up to 96% for the blends containing castor oil as SVO. For B20–B80 D/EA/SO triple blends, the opacity is reduced from 16 to 80% of the total opacity value attained with fossil diesel, Figure 5a. For its part, the reduction is even higher when B20–B80 D/EA/CO triple blends are employed, up to 94% lower than opacity obtained with diesel for B80 blend, Figure 5b. The slightly better behavior obtained with CO blends can be attributed to the lower presence of unsaturation that ricinoleic acid of

CO exhibit, in comparison to that exhibit by linoleic acid of SO. As it is well-known, unsaturation in fuels contributes to the formation of soot precursor species [36]. The results are in agreement with previous studies where triple mixtures with other oxygenated compounds reported a similar behavior in term of emissions reduction [6,16,17].

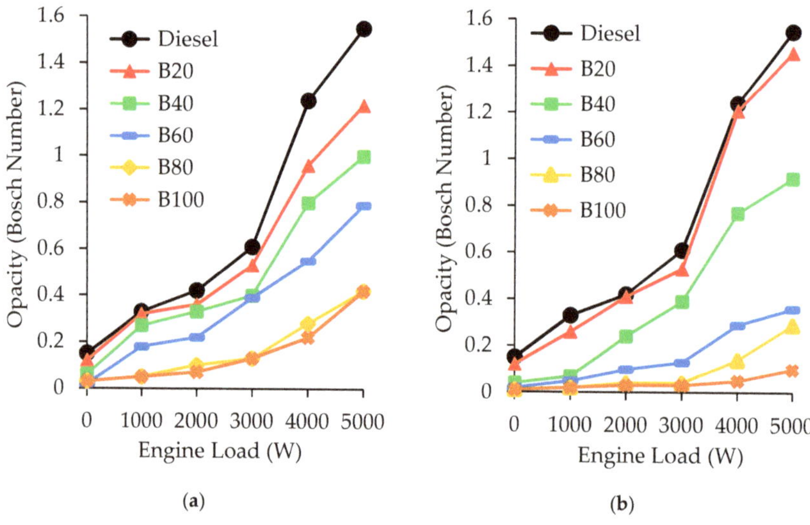


**Figure 5.** Smoke emissions (Bosch number) at different engine loads (1000–5000 kW) for (**a**) diesel/ethyl acetate/sunflower oil; (**b**) diesel/ethyl acetate/castor oil blends.

### 3.4. Fuel Consumption

The fuel consumption is also an important parameter to commercialize a new (bio)fuel. Figure 6 shows the influence of the different EA/SVO and D/EA/SVO blends with SO (Figure 6a) and CO (Figure 6b), on consumed volume (in liters per hour) by a diesel engine at low (1000 W), medium (3000 W), and high engine loads (5000 W). As biofuel ratio added to blends is rising, from B20 to B100, the volume consumed by the engine is greater. This fact is attributed to the fact that ethyl acetate has a lower calorific value than diesel (Table 1), which results in a reduction in the energy content of the mixtures, and therefore it is necessary to introduce more fuel on the engine to reach the required output power. As all the studied mixtures have lower calorific values than those of diesel, the engine consumes more fuel. The presence of sunflower oil leads to slightly higher fuel consumption compared to the analogous mixture that it contains CO, probably due to the lower cetane number of sunflower oil that increases the ignition delay, which deteriorates combustion process and leads to a higher fuel consumption. Particularly, blends B20–B80 with SO consume between 14 and 24% more than diesel, whereas the consumption of same blends with CO is about 4–21% more than diesel. On the other hand, as it is expected, the B100 blends display the highest percentage of consumption, 33% and 29% for EA/SO and EA/CO, respectively. For all the blends tested, the highest consumption as compared to diesel is usually produced at the highest engine load (5000 W).

To sum up, blends with ethyl acetate as LVS of sunflower and castor oils here evaluated has showed a greater efficiency on C.I. diesel engine than analogous mixtures tested in previous researches [16,17]. Triple blends containing DEE as renewable solvent achieved the best result with a proportion 60/18/22 D/DEE/CO, which led to a maximum of 40% of diesel replacement, and up to 77% of smoke emissions reduction [16]. Likewise, the same ratio in a blend containing ACE [17] gave rise to an equal percentage of fossil fuel substitution, but with emissions slightly lower (82%). In these cases, the engine does not work with higher amounts of DEE and ACE due to knocking problems attributed to the low energetic content of blends. Herein, the use of ethyl acetate allows to replace up to 100% of fossil diesel, with up to 94% in soot reduction respect to diesel. This means that all blends proposed can run in a conventional diesel engine, although it should be taking into account that a loss in engine power is produced as the concentration of pure biofuel EA/SVO supplied to the engine is increased.

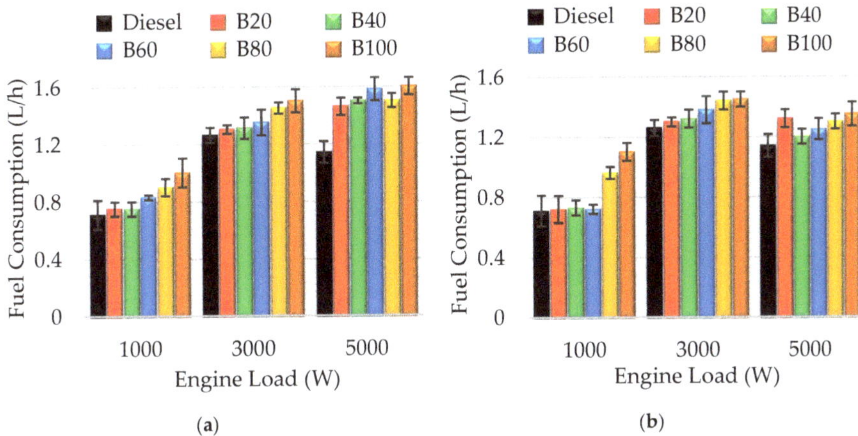


**Figure 6.** Fuel consumption values (L/h) at different engine loads (1, 3, and 5 kW) for (**a**) diesel/ethyl acetate/sunflower oil; (**b**) diesel/ethyl acetate/castor oil blends. The results are the average of three measurements and errors are represented as standard deviation through error bars.

## 4. Conclusions

In view of the current need for alternate fuels to supply the growing energy demand and also comply with new environmental requirements related to its renewable character, the objective of this research was to evaluate the potential of ethyl acetate as renewable biofuel in double blends with two different straight vegetable oils, castor or sunflower oil, and also, in triple blends along with fossil diesel. For that, the influence of these fuel blends on efficiency and emissions of a C.I. diesel engine was investigated. Considering the experimental results, the following points are concluded:

1. All investigated blends comply with requirements of kinematic viscosity (2.0–4.5 cSt) established by the European diesel standard EN 590, for usage in current diesel engines. However, calorific value and cetane number were reduced by the incorporation of ethyl acetate.
2. The addition of ethyl acetate led to remarkable improvements in flow properties of fuels at low temperatures in comparison to conventional diesel, which makes these fuels more suitable for engine running in colder climates.
3. Excellent results of power output were achieved with the B20–B60 D/EA/SO and with the B20–B80 D/EA/CO blends. The best performance was shown by fuels composed of castor oil, especially the B20 and B40 mixtures, exhibiting even better results than diesel at high engine load (4000 and 5000 W).
4. The consumption of engine fueled with the studied blends was greater than with conventional diesel due to low calorific value of ethyl acetate.
5. The high oxygen content of ethyl acetate was the key factor to enhance the combustion process and to achieve a very notable reduction of soot emissions. Higher proportion of EA/SVO in mixtures decrease the opacity generated by the engine. Hence, the mixtures B100 provided the better behaviour in term of achieving lower emissions.
6. This study also reveals that the EA/SVO double blends can be employed as direct biofuels without adding fossil diesel, making these fuels completely renewable.

7. The best results were achieved over the B40 blend with CO, which generated similar or even higher power output than diesel at the highest engine load values, with lower soot emissions and very similar fuel consumption. Moreover, this blend exhibited the best CP and PP values. With the B40 blend, a 60% of fossil diesel substitution was achieved. Furthermore, the production of ethyl acetate through renewable process like acidogenic fermentation from low-value biomass, constitutes a fundamental tool towards more sustainable production of alternative fuels for transportation sector.
8. Finally, the multi-component blending is a promising strategy to attain higher percentages of fossil fuel substitution, at the same time that exhaust emissions from the transportation sector are significantly diminished, keeping very good engine performance. This initial experimental study represents an advance in the search for new biofuels that can replace the fossil fuels used in the present fleet of vehicles.

**Author Contributions:** D.L. designed the study; L.A.-D. performed the experiments and wrote the paper; R.E. reviewed and edited the paper; R.E., D.L., and F.M.B. supervised the study and revised the manuscript; C.L., J.H.-C., J.C., A.P., and A.A.R. made intellectual contributions to this study. All authors have read and agreed to the published version of the manuscript.

**Funding:** This research received no external funding.

**Acknowledgments:** The authors are thankful to MEIC funds (Project ENE 2016-81013-R), to Andalusian Government (UCO-FEDER Project CATOLIVAL, ref. 1264113-R, 2018 call) and to Spanish MICIIN (Project ref. PID2019-104953RB-100), that cover the costs to publish in open access.

**Conflicts of Interest:** The authors declare no conflict of interest.

## Nomenclature

| | |
|---|---|
| ACE | Acetone |
| ASTM | American society for testing and materials |
| B0 | 100% diesel |
| B20 | 80% diesel + 20% EA/SVO blend |
| B40 | 60% diesel + 40% EA/SVO blend |
| B60 | 40% diesel + 60% EA/SVO blend |
| B80 | 20% diesel + 80% EA/SVO blend |
| B100 | 100% EA/SVO blend |
| BTE | Brake thermal efficiency |
| BSFC | Brake specific fuel consumption |
| CI | Compression ignition |
| CN | Cetane number |
| CO | Castor oil |
| CP | Cloud point |
| cSt | Centistokes |
| CV | Calorific value |
| D | Diesel |
| DEE | Diethyl ether |
| ISO | International Standards Organization |
| LVLC | Lower viscosity and lower calorific value |
| LVS | Low viscosity solvent |
| PP | Pour point |
| rpm | Round per minute ($min^{-1}$) |
| SO | Sunflower oil |
| SVO | Straight vegetable oil |
| VO | Vegetable oil |
| W | Watts |

**Symbols**

| | |
|---|---|
| A | Amperage (amps) |
| C | Calibration constant ($mm^2/s$)/s |
| P | Electrical power (watts) |
| t | Flow time (s) |
| V | Voltage (volts) |
| $\upsilon$ | Viscosity (centistokes) |

## References


1. Alalwana, H.A.; Alminshid, A.H.; Aljaafaric, H.A.S. Promising evolution of biofuel generations. Subject review. *Renew. Energy Focus* **2019**, *28*, 127–139. [CrossRef]
2. Rodionova, M.V.; Poudyal, R.S.; Tiwari, I.; Voloshin, R.A.; Zharmukhamedov, S.K.; Nam, H.G.; Zayadan, B.K.; Bruce, B.D.; Hou, H.J.M.; Allakhverdiev, S.I. Biofuel production: Challenges and opportunities. Review article. *Int. J. Hydrogen Energy* **2017**, *42*, 8450–8461. [CrossRef]
3. Estevez, R.; Aguado-Deblas, L.; Bautista, F.M.; Luna, D.; Luna, C.; Calero, J.; Posadillo, A.; Romero, A.A. Biodiesel at the Crossroads: A critical review. *Catalysts* **2019**, *9*, 1033. [CrossRef]
4. Luque, R.; Herrero-Davila, L.; Campelo, J.M.; Clark, J.H.; Hidalgo, J.M.; Luna, D.; Marinas, J.M.; Romero, A.A. Biofuels: A technological perspective. *Energy Environ. Sci.* **2008**, *1*, 542–564. [CrossRef]
5. Sivalakshmi, S.; Balusamy, T. Performance and emission characteristics of a diesel engine fueled by neem oil blended with alcohols. *Int. J. Ambient Energy* **2011**, *32*, 170–178. [CrossRef]
6. Hurtado, B.; Posadillo, A.; Luna, D.; Bautista, F.; Hidalgo, J.; Luna, C.; Calero, J.; Romero, A.; Estevez, R. Synthesis, performance and emission quality assessment of ecodiesel from castor oil in diesel/biofuel/alcohol triple blends in a diesel engine. *Catalysts* **2019**, *9*, 40. [CrossRef]
7. Rakopoulos, D.C. Combustion and emissions of cottonseed oil and its bio-diesel in blends with either n butanol or diethyl ether in HSDI diesel engine. *Fuel* **2013**, *105*, 603–613. [CrossRef]
8. Kumar, N.; Bansal, S.; Pali, H.S. *Blending of Higher Alcohols with Vegetable Oil-Based Fuels for Use in Compression Ignition Engine*; SAE Technical Paper 2015-01-0958; SAE International: Warrendale, PA, USA, 2015.
9. Kasiraman, G.; Nagalingam, B.; Balakrishnan, M. Performance, emission and combustion improvements in a direct injection diesel engine using cashew nut shell oil as fuel with camphor oil blending. *Energy* **2012**, *47*, 116–124. [CrossRef]
10. Kommana, S.; Naik Banoth, B.; Radha Kadavakollu, K. Eucalyptus-Palm kernel oil blends: A complete elimination of diesel in a 4-stroke VCR diesel engine. *J. Combust.* **2015**, *2015*, 182879–182886. [CrossRef]
11. Prakash, T.; Geo, V.E.; Martin, L.J.; Nagalingam, B. Evaluation of pine oil blending to improve the combustion of high viscous (castor oil) biofuel compared to castor oil biodiesel in a CI engine. *Heat Mass Transf.* **2019**, *55*, 1491–1501. [CrossRef]
12. Martin, M.L.J.; Geo, V.E.; Singh, D.K.J.; Nagalingam, B. A comparative analysis of different methods to improve the performance of cotton seed oil fuelled diesel engine. *Fuel* **2012**, *102*, 372–378. [CrossRef]
13. Vallinayagam, R.; Vedharaj, S.; Yang, W.; Roberts, W.L.; Dibble, R.W. Feasibility of using less viscous and lower cetane (LVLC) fuels in a diesel engine: A review. *Renew. Sustain. Energy Rev.* **2015**, *51*, 1166–1190. [CrossRef]
14. Lakshminarayanan, A.; Olsen, D.B.; Cabot, P.E. Performance and emission evaluation of triglyceride-gasoline blends in agricultural compression ignition engines. *Appl. Eng. Agric.* **2014**, *30*, 523–534.
15. Estevez, R.; Aguado-Deblas, L.; Posadillo, A.; Hurtado, B.; Bautista, F.; Hidalgo, J.M.; Luna, C.; Calero, J.; Romero, A.; Luna, D. Performance and emission quality assessment in a diesel engine of straight castor and sunflower vegetable oils, in diesel/gasoline/oil triple blends. *Energies* **2019**, *12*, 2181. [CrossRef]
16. Aguado-Deblas, L.; Hidalgo-Carrillo, J.; Bautista, F.; Luna, D.; Luna, C.; Calero, J.; Posadillo, A.; Romero, A.; Estevez, R. Diethyl ether as an oxygenated additive for fossil diesel/vegetable oil blends: Evaluation of performance and emission quality of triple blends on a diesel engine. *Energies* **2020**, *13*, 1542. [CrossRef]
17. Aguado-Deblas, L.; Hidalgo-Carrillo, J.; Bautista, F.; Luna, D.; Luna, C.; Calero, J.; Posadillo, A.; Romero, A.; Estevez, R. Acetone prospect as an additive to allow the use of castor and sunflower oils as drop-in biofuels in diesel/acetone/vegetable oil triple blends for application in diesel engines. *Molecules* **2020**, *25*, 2935. [CrossRef]

18. Kumar, A.; Rajan, K.; Kumar, K.R.S.; Maiyappan, K.; Rasheed, U.T. Green fuel utilization for diesel engine, combustion and emission analysis fuelled with CNSO diesel blends with Diethyl ether as additive. *Mater. Sci. Eng.* **2017**, *197*, 1–10. [CrossRef]
19. Krishnamoorthi, M.; Malayalamurthi, R. Availability analysis, performance, combustion and emission behavior of Bael Oil—Diesel—Diethyl ether blends in a variable compression ratio diesel engine. *Renew. Energy* **2018**, *119*, 235–252. [CrossRef]
20. Gurav, H.; Bokade, V.V. Synthesis of ethyl acetate by esterification of acetic acid with ethanol over a heteropolyacid on montmorillonite K10. *J. Nat. Gas Chem.* **2010**, *19*, 161–164. [CrossRef]
21. Sato, A.G.; Biancolli, A.L.G.; Paganin, V.-A.; da Silva, G.C.; Cruz, G.; dos Santos, A.M.; Ticianelli, E.A. Potential applications of the hydrogen and the high energy biofuel blend produced by ethanol dehydrogenation on a $Cu/ZrO_2$ catalyst. *Int. J. Hydrogen Energy* **2015**, *40*, 14716–14722. [CrossRef]
22. Contino, F.; Foucher, F.; Mounaïm-Rousselle, C.; Jeanmart, H. Experimental characterization of Ethyl acetate, Ethyl propionate, and Ethyl butanoate in a homogeneous charge compression ignition engine. *Energy Fuels* **2011**, *25*, 998–1003. [CrossRef]
23. Jones, R. Ethyl Acetate as Fuel or Fuel Additive. U.S. Patent Application No. 12,792,867, 8 December 2011.
24. Gangwar, J.N.; Saraswati, S.; Agarwal, S. Performance and emission improvement analysis of CI engine using various additive based diesel fuel. In Proceedings of the International Conference on Advances in Mechanical, Industrial, Automation and Management Systems (AMIAMS), Allahabad, Uttar Pradesh, India, 3–5 February 2017; pp. 189–195.
25. Wu, S.; Yang, H.; Hu, J.; Shen, D.; Zhang, H.; Xiao, R. The miscibility of hydrogenated bio-oil with diesel and its applicability test in diesel engine: A surrogate (ethylene glycol) study. *Fuel Process. Technol.* **2017**, *161*, 162–168. [CrossRef]
26. Zhang, L.; Shen, C.; Liu, R. GC–MS and FT-IR analysis of the bio-oil with addition of ethyl acetate during storage. *Front. Energy Res.* **2014**, *2*, 3. [CrossRef]
27. Çakmak, A.; Kapusuz, M.; Özcan, H. Experimental research on ethyl acetate as novel oxygenated fuel in the spark-ignition (SI) engine. In *Energy Sources, Part A: Recovery, Utilization, and Environmental Effects*; Taylor & Francis: Abingdon, UK, 2020.
28. Ashok, M.P. Study of the performance and emissions of the compression-ignition (CI) engine using ethyl acetate as a surfactant in ethanol-based emulsified fuel. *Energy Fuels* **2010**, *24*, 1822–1828. [CrossRef]
29. Ramadhas, A.S.; Jayaraj, S.; Muraleedharan, C. Biodiesel production from high FFA rubber seed oil. *Fuel* **2005**, *84*, 335–340. [CrossRef]
30. Keera, S.T.; El Sabagh, S.M.; Taman, A.R. Castor oil biodiesel production and optimization. *Egpt. J. Petrol.* **2018**, *27*, 979–984. [CrossRef]
31. Shahir, S.A.; Masjuki, H.H.; Kalam, M.A.; Imran, A.; Fattah, I.M.R.; Sanjid, A. Feasibility of diesel–biodiesel–ethanol/bioethanol blend as existing CI engine fuel: An assessment of properties, material compatibility, safety and combustion. *Renew. Sustain. Energy Rev.* **2014**, *32*, 379–395. [CrossRef]
32. Dwivedi, G.; Sharma, M.P. Impact of cold flow properties of biodiesel on engine performance. *Renew. Sustain. Energy Rev.* **2014**, *31*, 650–656. [CrossRef]
33. Krishnamoorthi, M.; Malayalamurthi, R. Experimental investigation on performance, emission behavior and exergy analysis of a variable compression ratio engine fueled with diesel-aegle marmelos oil-diethyl ether blends. *Energy* **2017**, *128*, 312–328. [CrossRef]
34. Labeckas, G.; Slavinskas, S.; Kanapkiene, I. The individual effects of cetane number, oxygen content or fuel properties on the ignition delay, combustion characteristics, and cyclic variation of a turbocharged CRDI diesel engine—Part 1. *Energy Convers. Manag.* **2017**, *148*, 1003–1027. [CrossRef]
35. Hellier, P.; Ladommatos, N.; Allan, R.; Payne, M.; Rogerson, J. The impact of saturated and unsaturated fuel molecules on diesel combustion and exhaust emissions. *SAE Int. J. Fuels Lubr.* **2012**, *5*, 106–122. [CrossRef]
36. Sendzikiene, E.; Makareviciene, V.; Janulis, P. Influence of composition of fatty acid methyl esters on smoke opacity and amount of polycyclic aromatic hydrocarbons in engine emissions. *Pol. J. Environ. Stud.* **2007**, *16*, 259–265.

MDPI
St. Alban-Anlage 66
4052 Basel
Switzerland
Tel. +41 61 683 77 34
Fax +41 61 302 89 18
www.mdpi.com

*Energies* Editorial Office
E-mail: energies@mdpi.com
www.mdpi.com/journal/energies

www.ingramcontent.com/pod-product-compliance
Lightning Source LLC
LaVergne TN
LVHW070649170726
843515LV00004B/362

* 9 7 8 3 0 3 6 5 0 2 7 8 6 *